D0102195

105

~BIBLIOTHECA•SCHOLÆ~
MERCATORVM•SCISSORVM
CROSBEIÆ•MAGNÆ•IN•AGRO
LANCASTRENSI•AB•IOANNE
~HARRISONO•EXSTRVCTÆ~
~ • • • • MDCXX • • • • ~

CONCORDIA • PARVÆ • RES • CRESCUNT.

Merchant Taylors' School Library

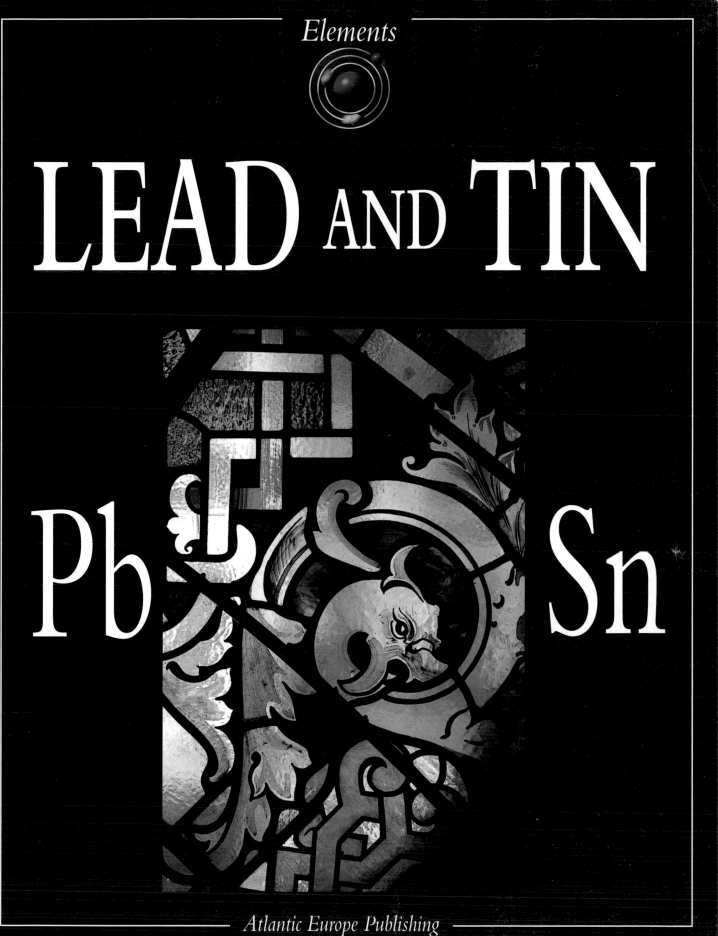

Elements

LEAD AND TIN

Pb

Sn

Atlantic Europe Publishing

How to use this book

This book has been carefully developed to help you understand the chemistry of the elements. In it you will find a systematic and comprehensive coverage of the basic qualities of each element. Each two-page entry contains information at various levels of technical content and language, along with definitions of useful technical terms, as shown in the thumbnail diagram to the right. There is a comprehensive glossary of technical terms at the back of the book, along with an extensive index, key facts, an explanation of the Periodic Table, and a description of how to interpret chemical equations.

The main text follows the sequence of information in the book and summarises the concepts presented on the two pages.

Technical definitions.

Substatements flesh out the ideas in the main text with more fact and specific explanation.

Equations are written as symbols and sometimes given as "ball-and-stick" diagrams – see page 48.

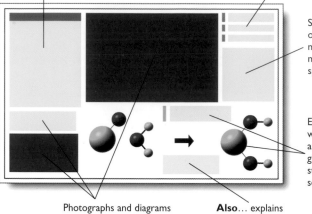

Photographs and diagrams have been carefully selected and annotated for clarity.

Also… explains advanced concepts.

An Atlantic Europe Publishing Book

Author
Brian Knapp, BSc, PhD
Project consultant
Keith B. Walshaw, MA, BSc, DPhil
(Head of Chemistry, Leighton Park School)
Industrial consultant
Jack Brettle, BSc, PhD (Chief Research Scientist, Pilkington plc)
Art Director
Duncan McCrae, BSc
Editor
Elizabeth Walker, BA
Special photography
Ian Gledhill
Illustrations
David Woodroffe
Designed and produced by
EARTHSCAPE EDITIONS
Print consultants
Landmark Production Consultants Ltd
Reproduced by
Leo Reprographics
Printed and bound by
Paramount Printing Company Ltd

Suggested cataloguing location
Knapp, Brian
    Lead and tin
    ISBN 1 869860 39 X
    – *Elements* series
540

Acknowledgements
The publishers would like to thank the following for their kind help and advice: *Dr Angus McCrae, Charles Schotman and Jonathan Frankel of* J.M. Frankel and Associates.

Picture credits
All photographs are from the **Earthscape Editions** photolibrary except the following:
(c=centre t=top b=bottom l=left r=right)
Mary Evans Picture Library 17, 31; **Tony Waltham, Geophotos** 12–13; **UKAEA** 27 and **ZEFA** 6, 36–37.

Front cover: Lead monoxide being reduced to molten lead by heating in the presence of carbon in the form of a block.
Title page: Because lead is malleable and unreactive, it is ideal for holding a complex pattern of irregularly shaped glass pieces together to make this beautiful leaded window.

First published in 1996 by
Atlantic Europe Publishing Company Limited, Greys Court Farm,
Greys Court, Henley-on-Thames, Oxon, RG9 4PG, UK.

Copyright © 1996
Atlantic Europe Publishing Company Limited
Reprinted in 1997

All rights reserved. No part of this publication may be reproduced, stored in a retrieval system, or transmitted in any form or by any means, electronic, mechanical, photocopying, recording or otherwise, without prior permission of the Publisher.

This product is manufactured from sustainable managed forests. For every tree cut down at least one more is planted.

The demonstrations described or illustrated in this book are not for replication. The Publisher cannot accept any responsibility for any accidents or injuries that may result from conducting the experiments described or illustrated in this book.

Contents

Introduction

An element is a substance that cannot be broken down into a simpler substance by any known means. Each of the 92 naturally occurring elements is therefore one of the fundamental materials from which everything in the Universe is made. This book is about the elements lead and tin.

Lead

You have probably noticed the word lead most commonly in petrol stations. Here much of the fuel is called lead-free, or unleaded. This tells us two important things about lead: it is very useful (otherwise lead would not have been put in petrol), and it has unpleasant side effects (otherwise why would such a fuss be made of taking lead *out* of petrol?).

In fact, people have always had a love-hate relationship with this important and commonly used metal. For example, lead was used by ancient peoples to make colouring pigments (often white) for paints. More disastrously, they also used lead in cosmetics. From this kind of use lead was absorbed into the body, where it acted as a poison. Lead in petrol could also have a similar effect, which is why it is now not used in fuels. But lead is also waterproof, easy to shape and does not corrode. It was just the thing for making water pipes – until it was discovered that lead dissolves in soft water and so contaminates the water.

The chemical symbol for lead is Pb. This is derived from *plumbum*, the Latin word for lead.

▲ Snail-eating utensils cast in pewter.

Perhaps the most confusing thing about this soft grey metal is that one of the most common uses we can think of for lead – for pencils – is not actually lead. Pencil "lead" is made from a form of carbon called graphite. In fact, if a pencil contained real lead, it would be very heavy to pick up, because lead is one of the heaviest metals on Earth.

Tin

The chemical symbol for tin is Sn, which comes from the Latin word *stannum*. It has many properties similar to lead; for example, it is a soft, weak metal and unreactive like lead.

Tin is found associated with copper and, like zinc, it has a low melting point. Tin was probably first used by ancient civilisations accidentally combined with copper. In this way they discovered bronze, the metal that gave its name to the period of ancient history called the Bronze Age.

Tin metal quickly develops an invisible protective coating when exposed to air. For this reason it has been used as a plating for protecting metals such as iron. But, because tin is less reactive than iron, tin is not an ideal material for protecting steel, and tin plate has been replaced by aluminium in many canning applications.

Tin is mainly found in alloys, including solders used in electronic circuits. Its other important role is as a liquid bed for the production of the sheet glass used for window panes, known as float glass.

Lead minerals

Lead forms a natural compound with many other elements, but it is most commonly seen as crystals of lead sulphide, where it is known as the mineral galena. It is a very heavy mineral, about three times as dense as most rocks.

Galena forms crystals that are shaped like cubes with a dull grey (lead grey) colour. The crystals are soft for a mineral, about the same hardness as a fingernail. This is why galena will mark surfaces with a dull grey line.

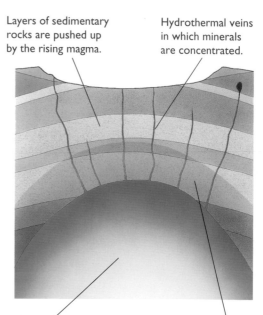

▼ The location of hydrothermal deposits in fissures above a magma source.

Layers of sedimentary rocks are pushed up by the rising magma.

Hydrothermal veins in which minerals are concentrated.

Magma from below the Earth's crust initially heats the surrounding rocks but eventually cools to form granite.

Rocks around the hot magma chamber are metamorphosed or changed.

▼ Lead ore being mined. Here a drill is being used to make holes for explosives to blast the tunnel face. The tunnel is following the metamorphosed rock that surrounds a fissure.

Where lead is found

Lead is widespread, occurring in the same location as other metals such as gold, silver and zinc. It mainly occurs as the mineral galena (lead sulphide, PbS), and also as cerussite (lead carbonate, $PbCO_3$) and anglesite (lead sulphate, $PbSO_4$).

Most concentrated metal deposits are formed in hydrothermal veins, that is, fissures in rocks sited above magma chambers. As hot mineral-rich fluids rise from the magma chambers during the last stages of volcanic activity, the fluids cool and a number of minerals solidify, among them lead. For this reason, ore mines usually recover a variety of metals.

Lead mines are sometimes found in areas of limestones (especially a form called dolomite) that once had magma chambers below them. Limestones have many natural fissures through which hot minerals can easily flow.

density: the mass per unit volume (e.g. g/cc).

hydrothermal: a process in which hot water is involved. It is usually used in the context of rock formation because hot water and other fluids sent outwards from liquid magmas are important carriers of metals and the minerals that form gemstones.

magma: the molten rock that forms a balloon-shaped chamber in the rock below a volcano. It is fed by rock moving upwards from below the crust.

mineral: a solid substance made of just one element or chemical compound. Calcite is a mineral because it consists only of calcium carbonate, halite is a mineral because it contains only sodium chloride, quartz is a mineral because it consists of only silicon dioxide.

sulphide: a sulphur compound that contains no oxygen.

vein: a mineral deposit different from, and usually cutting across, the surrounding rocks. Most mineral and metal-bearing veins are deposits filling fractures. The veins were filled by hot, mineral-rich waters rising upwards from liquid volcanic magma. They are important sources of many metals, such as silver and gold, and also minerals such as gemstones. Veins are usually narrow, and were best suited to hand-mining. They are less exploited in the modern machine age.

◄ The cubic structure and dull grey colour of lead sulphide shows very clearly in this specimen.

▼ Lead metal is bluish, but it readily oxidises in the presence of air to a dull grey. In polluted air it turns dark grey to black.

Galena

Lead sulphide, or galena, is easily spotted by its dark grey cubic (box-shaped) crystals.

Galena is soft. When it is rubbed against a surface, it leaves a grey streak of colour, showing its lead content. Like other lead compounds, galena is very heavy.

Galena is the main source of lead for the world's industries.

Refining lead in the laboratory

Lead ore normally occurs as an oxide or a sulphide. On these pages you can see a laboratory demonstration of the principle of refining the ore. You can then compare this with the industrial processing shown on pages 10 and 11. In this demonstration an orange form of lead monoxide (lead II oxide, PbO) is used, known as litharge.

To produce lead from its ore, the ore is first reduced (its oxygen removed) by placing it in a hollowed-out carbon block. The hollowing is important because the molten lead metal needs to be contained as the demonstration progresses.

Lead has a low melting point, so this demonstration proceeds under the heat of an ordinary Bunsen flame.

❷▼ The carbon block reduces some of the oxide and forms a molten globule of lead metal. Colourless carbon monoxide gas is given off during this reaction.

❶▼ Lead monoxide powder is heated on a carbon block.

EQUATION: Reduction of lead monoxide

Lead monoxide + carbon ⇨ lead + carbon monoxide

$2PbO(s)$ + $2C(s)$ ⇨ $2Pb(s)$ + $2CO(g)$

❸▶ The molten globule cools to leave solid lead.

Molten lead globule

ore: a rock containing enough of a useful substance to make mining it worthwhile.

oxide: a compound that includes oxygen and one other element.

reduction: the removal of oxygen from a substance.

reducing agent: a substance that gives electrons to another substance. Carbon is a reducing agent when heated with lead oxide.

Processing lead ore

Most lead is recovered from lead sulphide ore. The first stage of processing involves the concentration of the ore and the removal of sulphur. The ore contains a variety of other elements, each of which has to be removed.

  Because of the various impurities, lead is refined in a number of stages. Each stage depends on the contrast between the chemical and physical properties of lead and those of the other elements.

Processing lead ore

Lead ore is concentrated by crushing it to a fine powder and placing it into a vat with about three times as much water that contains wetting and frothing chemicals.

  Air is introduced into the vat, causing the frothing agents to make bubbles. The chemicals prevent the metal from becoming wet and metal grains are caught up by air bubbles, float, and can be skimmed off.

  Flotation increases the concentration of metal to about three-fifths. At this concentration the ore can be roasted – a process called sintering – which removes any sulphur as sulphur dioxide gas, leaving lead oxide behind. The byproduct of the sintering operation, sulphur dioxide gas, is usually converted to sulphuric acid and sold.

  Next, the lead oxide is smelted with a supply of coke and limestone in a furnace. The coke is almost pure carbon, and the reaction reduces the lead oxide to lead and forms carbon monoxide gas. The limestone forms a slag with some other impurities, and the metal flows free. It is tapped off the furnace and made into ingots.

  The ingots make "base bullion". This is not pure lead but contains small amounts of other metals. These are separated out and recovered when the lead is purified by the Parkes process (described opposite).

▼ Stages in the Parkes process.

❶ Impure lead ingot is melted and allowed to cool until the copper can be skimmed off.

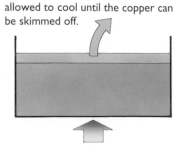

❷ The lead is reheated and air blown through it. More impurities oxidise, form a slag, and can be skimmed off.

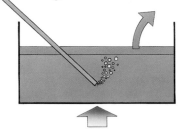

❸ Some zinc is added, and the alloy that forms contains zinc, silver and gold. These are skimmed off.

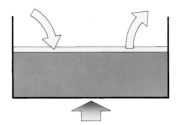

❹ The lead is finally purified by making it the anode of an electrolysis cell; pure lead collects at the cathode.

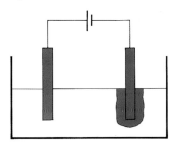

EQUATION: Removal of sulphur in lead sulphide ore by heating

Lead sulphide + oxygen ⇨ lead monoxide + sulphur dioxide

$$2PbS(s) \quad + \quad 3O_2(g) \quad \Rightarrow \quad 2PbO(s) \quad + \quad 2SO_2(g)$$

Refining lead by the Parkes process

In the Parkes process, contaminating elements are removed in a number of stages designed to allow the recovery of the impurities (which are themselves often valuable) as well as the lead.

Ingots of lead are melted and then allowed to cool. Copper has a higher melting point than lead, so as the mixture cools, crystals of copper float to the surface and can be skimmed off.

The lead is then reheated and a blast of air shot through it. This oxidises elements such as arsenic, and the oxides form a slag that can be skimmed off.

The next stage involves removing the silver and gold with a small amount of zinc. The silver and gold are more soluble than lead in zinc, so they alloy with the zinc. Zinc alloy is less dense that lead, so it rises to the surface and can be skimmed off. In this way zinc, silver and gold are all removed.

The remaining material, which is already nearly pure lead, can be further refined by electrolysis. An impure lead slab is used as the anode and a pure lead slab as the cathode. As the current flows, lead ions are attracted to the cathode of the cell and electroplated on it. When the cathode has accumulated sufficient lead, it is removed and replaced. The impurities on the anode consist mainly of valuable bismuth, and they, too, are collected.

alloy: a mixture of a metal and various other elements.

electrode: a conductor that forms one terminal of a cell.

electrolysis: an electrical–chemical process that uses an electric current to cause the break up of a compound and the movement of metal ions in a solution. The process happens in many natural situations (as for example in rusting) and is also commonly used in industry for purifying (refining) metals or for plating metal objects with a fine, even metal coating.

electrolyte: a solution that conducts electricity.

gangue: the unwanted material in an ore.

oxidation: a reaction in which the oxidising agent removes electrons. (Note that oxidising agents do not have to contain oxygen.)

pyrometallurgy: refining a metal from its ore using heat. A blast furnace or smelter is the main equipment used.

slag: a mixture of substances that are waste products of a furnace. Most slags are composed mainly of silicates.

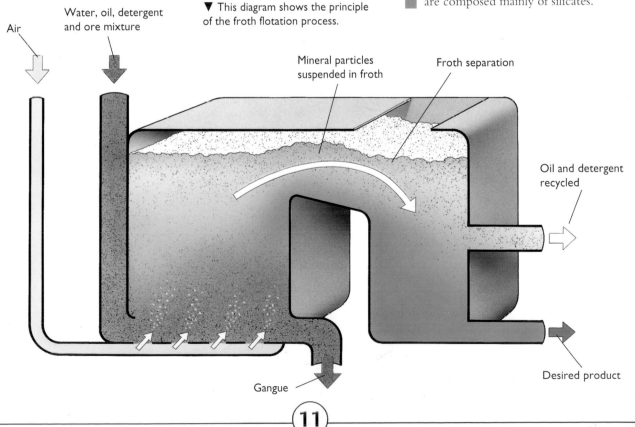

Air

Water, oil, detergent and ore mixture

▼ This diagram shows the principle of the froth flotation process.

Mineral particles suspended in froth

Froth separation

Oil and detergent recycled

Desired product

Gangue

Lead mine pollution

Lead has been mined throughout the world for centuries. However, until recently the process was not very efficient. For the most part, the ore was simply roasted to remove the sulphur, the sulphur dioxide being allowed to escape to the environment. As a result, lead smelting was a significant cause of air pollution and acid rain.

Furthermore, the gangue, which was simply piled up near to the mines, still contained considerable quantities of lead.

Lead compounds can be toxic to plants as well as animals. As a result, little will grow on many of the spoil heaps left from decades of inefficient mining.

Because lead dissolves very slowly in water, it will take a very long time for the lead to be washed from the upper part of the spoil heaps and for vegetation to recolonise the ground. Without positive conservation measures, such pollution will remain to scar the landscape for centuries.

◀ Leadville, Colorado, USA, was named after the lead deposits found at the site where the town grew up. It is in the Rocky Mountains, a site of former volcanic activity that is underlain by ancient granites.

The barren land in the foreground of this picture shows part of the spoil and also places where the ore was roasted. The area remains barren a century after most mining ceased.

The only way the land can be reclaimed is for the contaminated surface material to be removed and transferred to pits elsewhere. A start has recently been made on such reclamation using federal funds.

▼ Lead sulphide processing in China. This modern picture shows that much of the ore recovery is still done by simple roasting techniques, allowing the sulphur gases to escape to the environment, while the waste rock containing lead still accumulates over the landscape.

dissolve: to break down a substance in a solution without reacting.

toxic: poisonous enough to cause death.

Lead metal

Lead is the heaviest of all the common metals. It is over 11 times as heavy as an equal volume of water. It is a plentiful (and therefore cheap) material, whose unreactive properties make it useful for a wide range of applications. For example, it has been used for sculpture because it can be cast accurately.

The denseness of lead makes it a very useful material where compact weight is needed. You will find small pieces of lead sewn into the lower hem of some curtains to make them hang straight, and you will find lead in the bottom (keel) of a ship to balance the weight of the superstructure and to help stop it from rolling over. Lead, which is soft and will absorb shocks, can also be used as mounting blocks for heavy machines to help damp down vibrations.

Lead will also absorb radiation, and it has an important application in protecting radiographers and unexposed film from X-rays.

▲ Lead has a low melting point and can be used to make alloys suitable for fuse wires.

▼ Many lead compounds are insoluble. They are also heavy and readily settle out as soon as they have been precipitated. The tiny granules of the precipitate can be seen as they settle to the bottom of the test tube. The only common soluble lead compounds are lead nitrate and lead acetate.

Lead shot

For a long time small pieces of lead – called lead shot – were used by fishermen and clipped on to the line below a float to hold the float and tackle in the correct position in the water. This was banned after it was discovered that birds ate the shot and got lead poisoning.

For many centuries lead has also been used to make the shot that was loaded into guns. It was an ideal material, easily made molten, heavy and yet small, so it could easily travel through the air with the minimum loss of speed.

Lead shot is made by pouring molten lead from the top of a high tower. As the lead falls, it breaks up into tiny globules that solidify in the cool air. The lead reaches the bottom of the tower as round lead shot, where its fall is broken in a tub of water or oil.

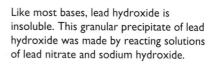

Like most bases, lead hydroxide is insoluble. This granular precipitate of lead hydroxide was made by reacting solutions of lead nitrate and sodium hydroxide.

density: the mass per unit volume (e.g. g/cc).

radiation: the exchange of energy with the surroundings through the transmission of waves or particles of energy. Radiation is a form of energy transfer that can happen through space; no intervening medium is required (as would be the case for conduction and convection).

X-rays: a form of very short wave radiation.

◀▲▶ Lead shot (small balls of lead) is widely used as weights in floats. In these examples the base of the float is weighted with lead shot so that the float stands upright in the liquid. In this way the float can be used as a hydrometer, a device for measuring the density of a liquid. The picture above shows the hydrometer in pure water; the one on the right shows it in salt water.

Lead ammunition

Lead is a suitable material for ammunition because it is heavy, cheap and easily formed into a shape. In this application its toxic chemical qualities are of no concern, for obvious reasons.

Lead-only bullets are made of lead alloyed with antimony. The cold alloy is sometimes forced through a die to make the correct bullet shape, or hot metal is poured into a mould. Lead deforms as it moves through the air, so lead bullets are used only in .22 calibre ammunition and in target-shooting handguns.

Ammunition used in high-velocity rifles has a lead–antimony core but is jacketed in brass or copper so that it keeps its shape.

Reactivity of lead

Many metals are subject to corrosion when placed in air or soil that is damp. The more reactive elements, such as sodium, cannot be used at all. Others that are quite reactive, such as iron, need to be protected by covering them in a protective coating of, say, paint.

Metal elements that are not very reactive can find many useful applications in weather–beaten environments, or where resistance to corrosion by water is needed.

All metal elements can be placed in order of their reactivity with other elements. The list of the more common metal elements is shown on this page. Those that are most reactive (and that can rarely be used in air or water) are at the top of the list, while those that can withstand exposure to air and water indefinitely are at the bottom of the list. Chemists call this list of elements the reactivity series.

You can see from the list on the right that only the precious metals (gold, silver, platinum) and mercury and copper are more resistant to corrosion than lead. The precious metals are too rare to have any wide use. This is why lead and copper have been so widely used for water systems and other locations exposed to possible corrosion. Lead has a low melting point, is cheap, and sheets can easily be formed into special shapes by hand, whereas copper is not as easily formed, is more expensive, and has to be beaten into shape.

REACTIVITY SERIES	
Element	Reactivity
potassium	most reactive
sodium	
calcium	
magnesium	
aluminium	
manganese	
chromium	
zinc	
iron	
cadmium	
tin	
lead	
copper	
mercury	
silver	
gold	
platinum	least reactive

▲ Lead is found towards the bottom of the reactivity series.

▶ Lead and lead-based alloys are frequently used as part of the protective sheathing for buried electricity cables. The lead is not only non-corrosive, but being soft, it will bend.

The lead alloy shown here is corrugated for extra strength and flexibility as it is extruded from the cable-making machine. Dense, soft lead will help to absorb damage that might be accidentally produced later by, for example, a mechanical digger reworking the site of the buried cable.

corrosion: the *slow* decay of a substance resulting from contact with gases and liquids in the environment. The term is often applied to metals. Rust is the corrosion of iron.

inert: nonreactive.

precious metal: silver, gold, platinum, iridium, and palladium. Each is prized for its rarity. This category is the equivalent of precious stones, or gemstones, for minerals.

pyrite: "mineral of fire". This name comes from the fact that pyrite (iron sulphide) will give off sparks if struck with a stone.

Lead containers

Why would you want to make a container out of one of the world's heaviest metals? The answer lies in a key property of lead: it will not corrode. So, for just the same reason as it was once useful for making water pipes, lead was traditionally used to make containers.

When a container is made from lead, the surface atoms of the lead react with the oxygen in the air to make lead oxide. When lead comes into contact with sulphuric acid, it makes an inert coating of lead sulphate.

Normally, it is very difficult to find containers for sulphuric acid because it dissolves them, but because lead reacts very slowly to form a protective coating of lead sulphate, sulphuric acid can readily be kept in lead. In fact you may well find a lead-lined tub of sulphuric acid in some school metal workshops, where the acid is used for removing the oxide from most metals.

Lead chamber process

The lead chamber process was the earliest method of preparing sulphuric acid. Its development was one of the landmarks in the development of industrial chemistry because it allowed a valuable raw material, sulphuric acid, to be produced in large quantities for the first time. This, in turn, allowed the acid to be produced much more economically than previously and so brought down the price of chemicals and the processes and products that depended on them.

The lead chambers were the size of rooms. In this process a mixture of sulphur or pyrite and potassium nitrate was placed in a ladle and ignited. The floor of the chamber was covered with water. As the mixture burned, the gases produced condensed on the inside of the lead-lined chamber and were absorbed by the water.

The process produced about two-thirds purity sulphuric acid. The purpose of the lead chambers was simply to act as an inert container (since the lead forms an insoluble sulphate coating when in contact with the sulphuric acid).

Lead oxides

Lead oxides produce a variety of colours. For example, the most important oxide of lead (called litharge) is an orange substance used in a wide variety of applications. A mixture of litharge and red lead is used as the active coating on the plates of all lead–acid batteries (see page 28). It can be used in glass-making to produce glass that will prevent the penetration of X–rays and other radioactive sources.

The distinctive colour of litharge is used in pottery glazes and enamels and as a toughener in rubber. Made into chrome yellow, it is incorporated into some paints.

Lead oxide is also important when dissolved in sodium hydroxide, where it acts to break up some sulphur compounds in petroleum refineries.

Red lead is used in some paints designed to prevent iron and steel from corroding (see page 24).

▶ Lead oxides make a variety of colours. Massicot is a yellow form of lead monoxide (lead II oxide or PbO). Lead monoxide is one of the most widely used and commercially important metallic compounds. The bright orange-red powder is known as "red lead" (Pb_3O_4), whereas the darkest powder is lead dioxide (lead IV oxide, PbO_2).

EQUATION: Oxidation of lead

Lead + oxygen ⇨ lead monoxide (litharge)

$$2Pb(s) \quad + \quad O_2(g) \quad ⇨ \quad 2PbO(s)$$

Red lead

Red lead is an example of a mixed oxide. Although its formula is Pb_3O_4, it behaves chemically as though it were a mixture of lead monoxide (lead II oxide, PbO) and lead dioxide (lead IV oxide, PbO_2). When it is heated, it decomposes to lead monoxide and releases oxygen.

EQUATION: Heating red lead to release oxygen

Red lead ⇨ lead monoxide + oxygen

$$2Pb_3O_4(s) \quad ⇨ \quad 6PbO(s) \quad + \quad O_2(g)$$

decompose: to break down a substance (for example by heat or with the aid of a catalyst) into simpler components. In such a chemical reaction only one substance is involved.

oxidation: a reaction in which the oxidising agent removes electrons. (Note that oxidising agents do not have to contain oxygen.)

oxide: a compound that includes oxygen and one other element.

Lead nitrate and sulphide

Lead is not a very reactive metal, and the compounds it forms are therefore not very stable. As a result it is relatively easy to get lead compounds to break down, or decompose.

Some lead compounds also have photoelectric properties, that is, they will change their conductivity when infra-red light reaches them.

Both of these properties are demonstrated on these pages.

Lead sulphide as a photoconductor

When some lead compounds absorb light, electrons are freed, making them more conductive. Lead sulphide (PbS), lead selenide (PbSe) and lead telluride (PbTe) are all sensitive to infra-red radiation (although they are not sensitive to visible light).

Each of these photo-conducting substances finds a role in detecting infra-red radiation. For example, photoconductors can detect the infra-red radiation given out by warm bodies. It is for this reason that they are used to make PIR (Passive infra-red) detectors, now commonplace as part of home security. They are used to trigger alarms or switch on lights when a warm body (such as a person or a vehicle) comes within their detection range.

▲ A PIR detector containing a lead sulphide photoconductor.

Also:

Cadmium sulphide is sensitive to changes in visible light, and it is the semi-conducting material used in dusk-to-dawn light switches and camera photocells.

Decomposition of lead nitrate

Lead nitrate decompose to lead oxide when it is heated. As the lead nitrate decomposes, energy is released, and the crystals of lead nitrate break up. The process of breaking down is called decrepitation; it produces a loud crackling sound.

Decrepitation releases both nitrogen dioxide and oxygen, the gases collected at the left hand end of the apparatus.

❶▼ The apparatus set up before the experiment

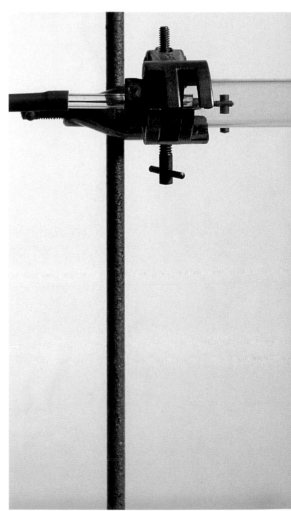

decompose: to break down a substance (for example by heat or with the aid of a catalyst) into simpler components. In such a chemical reaction only one substance is involved.

electron: a tiny, negatively charged particle that is part of an atom. The flow of electrons through a solid material such as a wire produces an electric current.

infra-red radiation: a form of light radiation where the wavelength of the waves is slightly longer than visible light. Most heat radiation is in the infra-red band.

❷◀ Lead nitrate is heated, releasing gases and producing crackling sounds.

Lead nitrate, a white powder, being heated. It changes colour as it melts and decomposes to lead oxide. The bubbles and the brown fumes are oxygen gas and nitrogen oxides.

❸▼ A close-up of the decomposition of lead nitrate.

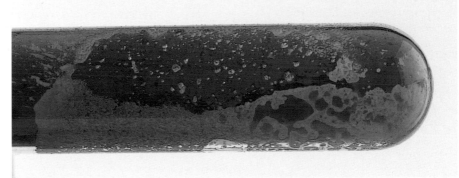

Bunsen burner

EQUATION: Decomposition of lead nitrate

Lead nitrate ⇨ lead monoxide + nitrogen dioxide + oxygen

$$2Pb(NO_3)_2(s) \quad ⇨ \quad 2PbO(s) \quad + \quad 4NO_2(g) \quad + \quad O_2(g)$$

Lead for buildings

Lead has many natural advantages for building. It has been widely used in the past and continues to be used today.

Lead's properties are ideal for building: it is waterproof, soft and easily moulded into awkward corners that cannot be filled by other materials. It is heavy, holding the roof of a building down even in high winds, and it does not corrode, so a lead roof will remain in good repair for many centuries.

In the Middle Ages, most cathedral roofs in Europe were made of sheet lead, which is why they have a dull grey appearance. The lead roof of a cathedral in Istanbul, Turkey, is known to have lasted without repair for over 1400 years.

Lead is also used for protecting buried cables and for sealing cast iron pipes in some sewage systems.

Lead in water systems

The word plumbing is derived from the Latin word for lead, *plumbum*. For many centuries people did not have any idea that lead was a danger to health. They could not detect the minute amounts of lead compounds that dissolved in water, and so people could only see the advantages of the metal.

One of the main needs was to get water from rivers or reservoirs to homes inside cities. The Romans were one of the first civilisations to achieve this, carrying water cross-country in long stone aqueducts. However, these open waterways were of no use in the city; instead, pipes were needed to bring water to homes. Some pipes were made of naturally hollow materials such as bamboo, or cast materials such as clay, but most pipes were moulded from lead because it was soft, did not corrode, and could be easily shaped to go round bends and other difficult areas of buildings. One pipe could also be joined to another very easily.

Although the Romans were the first to use them, lead water pipes continued to be used for most plumbing systems well into the present century. It was only quite recently that the possible dangers of lead in pipes became known. Today millions of homes are still supplied by water through lead pipes.

Also:

There are many, and sometimes confusing, uses of the word lead in connection with glass. The earliest use may be when, in about 1675, the English glassmaker Ravenscroft developed "glass of lead" which was more durable than the glass then being made. Lead crystal is glass containing enough lead to alter the refractive index, while leaded glass refers to the way glass panes are fixed with lead strips.

corrosion: the *slow* decay of a substance resulting from contact with gases and liquids in the environment. The term is often applied to metals.

◀ Lead was a common roofing material for grand buildings because it could readily be moulded into shape. Beginning with St Sofia in Instanbul, the fashion for leaded domes spread across Europe. This picture is of Venice.

Lead is used more sparingly on modern buildings, but it still has an important function as lead flashing. The soft lead sheet is let into the brick courses and then laid out and moulded over the top edge of a window. This provides an easily laid, long-lasting and waterproof seal for the top of the window.

Leaded glass

In the past it was only possible to make small panes of glass. As a result, some means of joining them together had to be found. This is why many old windows were made of panes of glass held together with strips of lead.

Lead is soft and was easily moulded around the panes. Lead also melts at temperatures far below that of glass and so could be melted to run into the window frames and seal the glass.

(Modern frames use diamond leading which is simply stuck on to the outside of the pane for decoration. It serves no other purpose.)

◀ Traditional diamond leaded windows as used in the original Maryland State House complex, Maryland, USA.

Lead in paint and glaze

A paint is a liquid that is applied to a surface to protect it and to give it a colour. A paint is made up of solid colouring material – known as a pigment – mixed into a liquid that allows the pigment to be applied with a brush or spray gun. When it is applied, the liquid evaporates or reacts chemically with the surface, leaving the pigment behind as a thin solid film.

One of the most common pigments used in paint was white lead; a general term for a range of white lead pigments such as lead carbonate, lead sulphate and lead silicate. For many years, it was used in interior gloss paints and in exterior paints; but because of the possibility that very young children might eat the paint, it is no longer used for interior or domestic purposes.

The other coloured oxides of lead have also been widely used. Red-leaded paints are still used for exterior applications because lead oxide and lead silicates are so stable they resist corrosion even in the most exposed places. Red lead is still widely used as a metal–protective undercoat.

▼ Red lead being applied as a protective coat after a rusty part has been stabilised with phosphoric acid.

The drying properties of lead

When a paint has been applied, it is still liquid, and the faster it can be made to dry, the more attractive it will be. The first phase of the drying occurs when the solvent used to apply the pigment evaporates.

But often a further chemical reaction is needed to make the paint into a hard solid. Many paints – gloss paints, for example – use an oil base. In these cases hardening ("drying") occurs as the oil polymerises on exposure to air, forming long chains of material that can no longer move about.

These setting processes are speeded up by the use of a catalyst. Several metals, including lead, have been widely used as drying additives.

catalyst: a substance that speeds up a chemical reaction but itself remains unaltered at the end of the reaction.

ceramic: a material based on clay minerals, which has been heated so that it has chemically hardened.

flux: a material used to make it easier for a liquid to flow. A flux dissolves metal oxides and so prevents a metal from oxidising while being heated.

pigment: any solid material used to give a liquid a colour.

Lead glaze

The shiny, smooth, glass-like finish on many pots and other ceramic objects is called a glaze. Colour and decoration are also added to the glaze, while the hard smooth finish makes the finished ceramic object much easier to wash and clean.

The glaze is made from powdered glass mixed with a colouring material. Lead oxide is a common colouring substance. Powdered glass and lead oxide are mixed in water and then coated on the pot or other object. The glaze is then dried and heated in a kiln to make the glass powder soften and form a continuous coating, and thus fix the glaze in place. At the same time, the glaze reacts chemically with the ceramic and so forms a permanent bond.

The lead oxide has two roles. It adds a white colour to the glaze, and at the same time, it lowers the temperature at which the glaze melts and ensures that it runs more smoothly while in the kiln. This property of a material makes it an efficient flux.

▶ Lead is a common component of glaze, the reflective surface coating on pottery.

Lead and glass

Glass is one of the most common materials we use. It is formed from sand with additional compounds such as soda (sodium carbonate), limestone and lead. It is heated to about 1300°C, when the reactants chemically combine to produce glass (and give off gaseous products).

The Venetians were the first to add lead to the glass-making mixture. The lead allows the glass to be moulded over a wider range of temperatures. This, in turn, allowed the Venetians to make glass in more complicated shapes.

However, the lead-containing glass called crystal was developed in England in the 17th century. The use of lead made the glass even clearer and therefore more transparent. Also, the lead changed the optical properties of the glass, giving cut glass a special sparkle. This helped to make London the foremost glass-making centre of the 18th century.

▲ Lead glass (crystal)

The effect of lead additives

Most glasses are a form of silica made from a network of silica molecules. In quartz these molecules are packed together in a regular framework; but when other atoms, such as sodium, calcium or lead, are added, the rigid framework is somewhat broken up. This is what makes glass less viscous than molten quartz, and why it also melts at lower temperatures.

Lead makes glass sparkle because it changes the optical properties of the glass, increasing the refractive index of the glass and thus increasing the reflectivity. A traditional lead crystal glass contains (by weight) about 56% silica, 30% lead oxide, 2% sodium oxide and 12% potassium oxide.

Optical glass, for example, glass used in lenses, has a broad spectrum of lead content, the amount varying between 3 and 70% depending on the refractive index required.

Lead also has the ability to absorb radiation, so optical glass with a high lead content (64% lead oxide, 27% silica) is used for remote viewing of X-ray facilities, hot cells in nuclear research, irradiation facilities for food preservation and medical sterilization installations.

▶ Lead glasses have about half the radiation absorption capability compared with lead sheet of the same thickness, which is why the glass is usually made thicker than the remaining lead shielding.

Solder glasses

A variety of lead glasses are important as low-melting sealing and solder glasses. A solder glass is a piece of glass used to join two pieces of glass together. This may be important, for example, in joining two intricate shapes or when sealing glasses to metals, as in the manufacture of television tubes.

For these purposes a low softening point glass is needed so that the solder glass can be softened and used at temperatures well below the softening points of the two pieces of glass that are to be joined. This glass will soften at about 630°C. A typical sealing glass contains about 23% lead oxide by weight. (Glass used for fine lead crystal contains about 30% lead oxide by weight.)

bond: chemical bonding is either a transfer or sharing of electrons by two or more atoms. There are a number of types of chemical bond, some very strong (such as covalent bonds), others weak (such as hydrogen bonds).

oxide: a compound that includes oxygen and one other element.

refractive index: the property of a transparent material that controls the angle at which total internal reflection will occur. The greater the refractive index, the more reflective the material will be.

viscous: slow moving, syrupy. A liquid that has a low viscosity is said to be mobile.

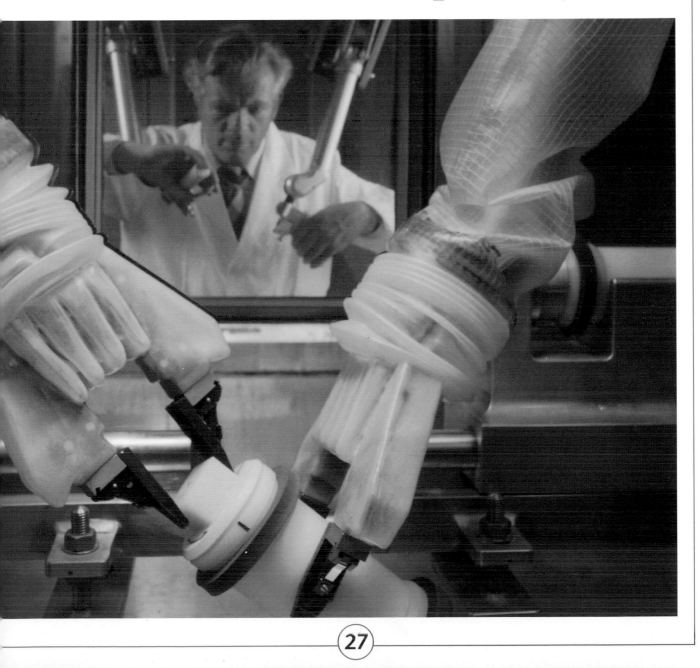

Lead in batteries

One of the biggest users of lead and lead oxide is the storage battery industry. The main plates used in lead–acid batteries are made from grids of lead alloy, alternate plates being coated with "spongy" lead and lead oxide.

The lead–acid battery is made of a number of units, each called cells. Every cell in a lead–acid battery can generate approximately 2 volts of electricity. In a car battery, six cells are placed in series, producing a total of about 12 volts.

Bubbles of gas seen during charging are due to the splitting of some water molecules into oxygen and hydrogen gases.

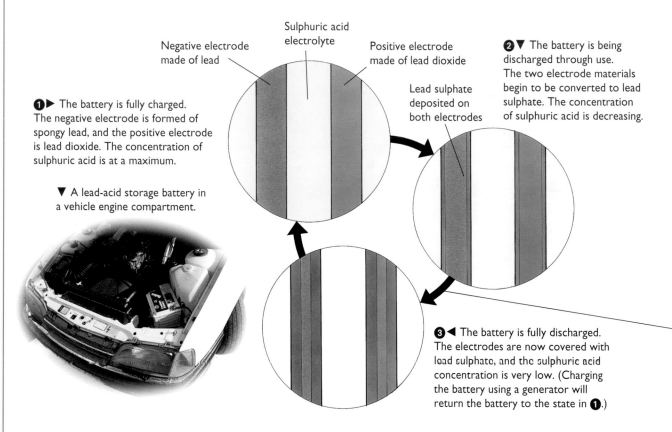

Sulphuric acid electrolyte

Negative electrode made of lead

Positive electrode made of lead dioxide

Lead sulphate deposited on both electrodes

❶ ▶ The battery is fully charged. The negative electrode is formed of spongy lead, and the positive electrode is lead dioxide. The concentration of sulphuric acid is at a maximum.

▼ A lead-acid storage battery in a vehicle engine compartment.

❷ ▼ The battery is being discharged through use. The two electrode materials begin to be converted to lead sulphate. The concentration of sulphuric acid is decreasing.

❸ ◀ The battery is fully discharged. The electrodes are now covered with lead sulphate, and the sulphuric acid concentration is very low. (Charging the battery using a generator will return the battery to the state in **❶**.)

EQUATION: Discharging (using) a lead-acid battery

Lead + lead dioxide + sulphuric acid ⇨ lead sulphate + water

$$Pb(s) + PbO_2(s) + 2H_2SO_4(aq) \Rightarrow 2PbSO_4(s) + 2H_2O(l)$$

EQUATION: Charging a lead-acid battery

Lead sulphate + water ⇨ lead + lead dioxide + sulphuric acid

$$2PbSO_4(s) + 2H_2O(l) \Rightarrow Pb(s) + PbO_2(s) + 2H_2SO_4(aq)$$

How lead-acid batteries work

Lead-acid batteries are the most commonly used form of vehicle battery. To work as an electrical battery, each of the cells inside it must have two electrodes made of electrically conducting materials. These must be bathed in a liquid that can conduct electricity as well as help the battery store electricity. This material is called an electrolyte.

The electrodes, or plates, in lead-acid batteries are designed to have a large surface area. This allows the charge stored as chemical energy to be converted quickly into electrical energy. Half of the electrodes are made from a lead alloy covered in a paste of lead dioxide; the other half are made simply of lead alloy (known as "spongy" lead). All the plates are bathed in dilute sulphuric acid.

When a charge is drawn from the battery, a chemical reaction occurs between the plates and the sulphate ions of the electrolyte that changes the lead dioxide into lead sulphate.

When the battery is charged (perhaps from a vehicle generator), and a current passed through the battery, a chemical reaction occurs that turns the lead sulphate on the plates into lead and lead dioxide and returns the sulphate to the electrolyte, thus increasing the concentration of the acid.

When a load is applied to the battery, the reaction reverses, the lead sulphate coatings re-form and the sulphuric acid is used up.

This process can be repeated many times, giving the secondary battery a useful life of many years of constant service.

battery: a series of electrochemical cells. A 12V battery contains 6 cells, each providing 2V.

cell: a vessel containing two electrodes and an electrolyte that can act as an electrical conductor.

current: an electric current is produced by a flow of electrons through a conducting solid or ions through a conducting liquid.

electrolyte: a solution that conducts electricity.

ion: an atom, or group of atoms, that has gained or lost one or more electrons and so developed an electrical charge. Ions behave differently from electrically neutral atoms and molecules. They can move in an electric field, and they can also bind strongly to solvent molecules such as water. Positively charged ions are called cations; negatively charged ions are called anions. Ions carry electrical current through solutions.

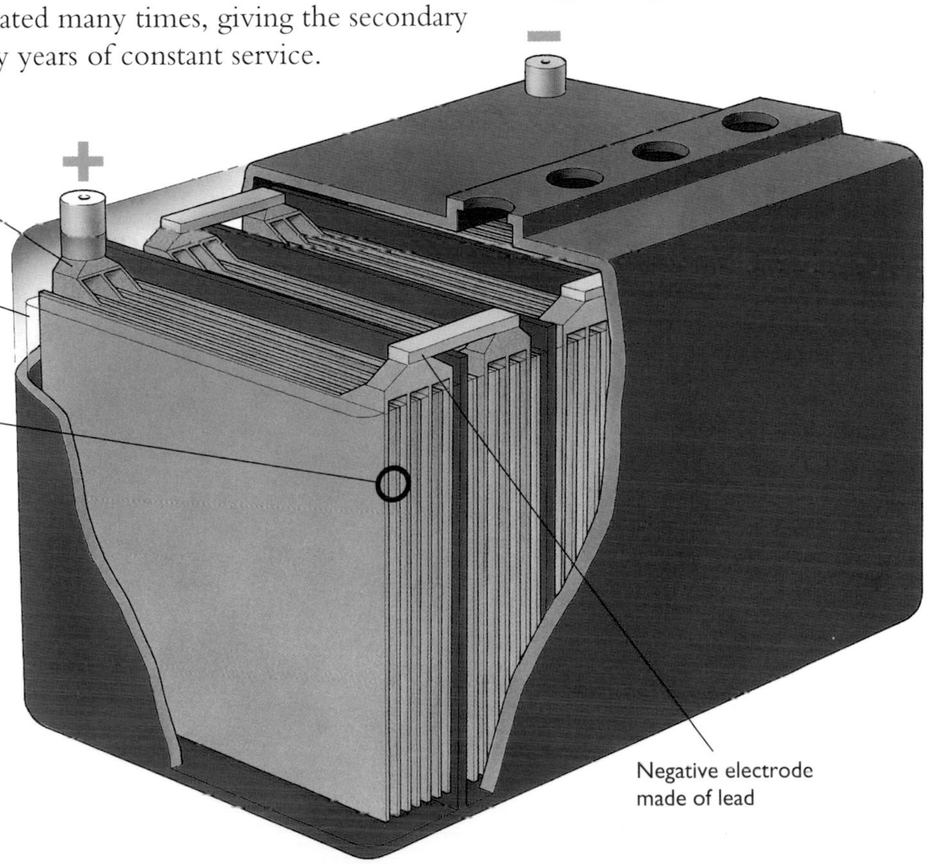

Positive electrode made of lead dioxide

Sulphuric acid electrolyte

▶ This diagram shows the arrangement of lead and lead oxide plates. Notice that each cell is connected in series to the next through conductors at the top of the plates.

Negative electrode made of lead

Lead and the body

Lead is a heavy, unreactive substance. These very properties have been used for the benefit of health, and they have also caused major health problems.

Although lead has many advantages outside the body, such as protecting people from overdoses of harmful radiation, it has become widely known in recent years for its harmful effects inside the body. Examples of this are given on these pages and on pages 32 and 33. Lead is by no means the only substance that can harm the body; but because it was used so widely in fuels and paints, its effects were readily noticed.

Lead in cosmetics of the past

Body painting was the earliest use of cosmetics, meant to protect the body from evil spirits. Many of the original natural cosmetics also helped to protect the wearer from flies. As a result, cosmetics were most often applied to the areas around the eyes. One of the most common cosmetics was a white pigment made from white lead. The ancient civilisations of the Middle East used to paint eyelashes and eyebrows with a black paste that included lead sulphide (galena) and soot. Later, the Romans made white face powder using white lead. It was extremely poisonous.

The modern art of make-up developed in Europe, especially in France and in Venice, Italy. Venetians made a specially desirable form of paste known as cerise. It had the effect of making the skin of the face white. It, too, was made from white lead.

People knew that using lead on the face would eventually result in lead poisoning, causing deterioration of complexion, baldness and eventually death. Nevertheless, so great was the desire to look beautiful that many poisoned themselves through this particularly dangerous form of make-up.

▶ In Roman times, and despite being known as a poison, lead was added to drinks because it enhanced their flavour.

Lead poisoning

From ancient times, pottery workers were known to suffer from lead poisoning because of breathing in lead fumes and handling lead glazes. At the beginning of the 20th century, lead was so widely used for industrial processes that researchers found 77 occupations at risk from lead poisoning. Such evidence began to make protection from harmful substances seem necessary.

Lead is absorbed very slowly. Unfortunately, the body has no natural way of getting rid of any lead that is absorbed, so any lead taken in as fumes or in food, or in liquids, simply builds up in the body, especially in the liver and kidneys.

People suffering from lead poisoning lose their appetite, start to be sick and may develop convulsions. In severe cases the brain is damaged and death may follow.

The most important sources of lead poisoning are thought to be caused by lead in some drinking water supplies (especially in soft-water areas) and in vehicle exhausts.

Studies show that children are particularly vulnerable to lead poisoning. This knowledge has resulted in the replacement of lead with copper water pipes, the removal of lead from domestic paints and the change from leaded to unleaded fuel.

EQUATION: How lead dissolves in water

Lead + soft water + dissolved oxygen ⇨ lead hydroxide

$$2Pb(s) \quad + \quad 2H_2O(l) \quad + \quad O_2(g) \quad \Rightarrow \quad 2Pb(OH)_2(aq)$$

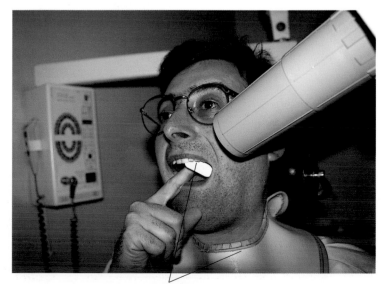

Lead shield

radiation: the exchange of energy with the surroundings through the transmission of waves or particles of energy. Radiation is a form of energy transfer that can happen through space; no intervening medium is required (as would be the case for conduction and convection).

radioactive: a material that emits radiation or particles from the nucleus of its atoms.

X-rays: a form of very short wave radiation.

Lead protects the body from radiation

X-rays, which are produced when charged particles collide, are a form of radiation whose waves have a much shorter wavelength than that of visible light.

How well X-rays can pass through a material depends on the density of the atoms of the material. Bones, for example, are very dense and block most radiation, and this is what causes their X-ray image to appear white. Tissue blocks less radiation and appears darker on the film. Lead is composed of large, closely packed atoms that will absorb all X-rays.

One common use of lead is therefore as a shield against radiation, for example, to protect those who make X-ray pictures in hospitals and dental surgeries. It is also used in nuclear power stations and in the containers used to transport radioactive materials.

◄ Lead is made of closely packed large and heavy atoms. This makes it very difficult for radiation to pass through it.

Lead in fuels

We use petrol in most of the world's engines. Petrol is made from refined crude oil, which is a compound of carbon. The fuel is "atomised" and sprayed into the engine cylinders along with air. The piston then rises in the cylinder and compresses the mixture, causing it to get hot.

As the air is being compressed, and before a spark is generated by the spark plug, the mixture may explode. This is called pre-ignition, and it is noticed as a knocking sound inside the engine.

The numbers used to describe the quality of the fuel are octane numbers. Commonly the octane number varies between 79 and 91. This number describes the way the fuel can resist "knocking".

When engines knock, it means that the mixture has exploded while the piston is still travelling up the cylinder, not when it has just passed the top of its stroke, as is meant to happen. By exploding at the wrong time, the engine become both noisy and less efficient, and so consumes more fuel.

Compounds of lead (tetraethyl lead, $Pb(C_2H_5)_4$) were traditionally added to the fuel to stop the mixture from exploding prematurely. This meant that engines ran more efficiently. This was a good example of how a simple chemical treatment provided a cheap solution to a problem. However, because it has now been realised that lead can be toxic in the environment, this use of lead is now being phased out. Instead, the engines have had to be redesigned so that they do not knock even when lead additives are not present.

atomised: broken up into a very fine mist. The term is used in connection with sprays and engine fuel systems.

mixture: a material that can be separated out into two or more substances using physical means.

octane: one of the substances contained in fuel.

Catalytic converters

These work by passing the spent exhaust fumes from the engine over a fine mesh coated with platinum or other material. This makes the gases turn into less poisonous forms. However, if a catalytic converter were used with leaded fuel, then the lead in the exhaust would coat the mesh and make the converter inoperable. This is why converters can only be used with lead-free fuel.

◀ The first country to change from leaded to unleaded fuel was the United States, following a ruling by the Environmental Protection Agency in 1985. Most other industrial world countries have followed the same pattern, so that it is now less common to find leaded fuel on sale.

▼ The oxidation of gases in internal combustion engines.

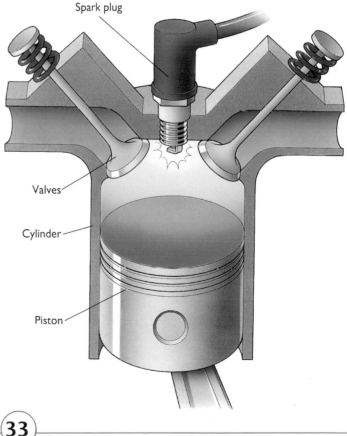

Spark plug

Valves

Cylinder

Piston

Tin

Tin is a relatively rare, silvery white, soft metal with a low melting point (232°C). It is mostly found as tin oxide, cassiterite, in veins that are close to igneous rocks (the same conditions that gave rise to lead deposits, see page 7). Tin compounds were probably deposited in these veins by hot fluids rising above molten magma.

Tin is a very soft and weak metal, so it is never used on its own. Tin is also very brittle; a bar of tin will break up into sharp crystals if bent, as well as making a high pitched sound like crying.

The main use of tin has traditionally been as a plating on steel (see page 38), utilising the noncorrosive properties of the metal.

However, tin has many other uses, especially as an alloy. Although tin is a weak metal on its own, an alloy that includes tin is typically much harder than one without it. There are other advantages, too, such as making the alloy easier to cast. Tin is found in some brass, bronze and pewter.

The low melting point of tin makes it a useful component of solder and for preparing alloys that have low melting points. For this reason, it can be used in fire-detection systems.

Ore deposits

Tin, like many other poorly reactive metals, such as tungsten, copper and silver, is formed in the metamorphic zones surrounding ancient magma chambers. It was present in the hydrothermal fluids that rose up from the magma chambers as they cooled during the last stages of volcanic activity. The fluids flowed into fissures that had been produced as the magma chambers domed up the overlying rocks. As the fluids rose in the fissures, they cooled, and a wide variety of minerals solidified as the temperature dropped below their various melting points. Tin-bearing lodes are normally found making a ring of fissures over a magma chamber.

The only lode deposits to have been found are in Cornwall, England and Bolivia. These rich lodes are mined and the tin processed by direct smelting. The type of mining involved is shown on these pages.

Elsewhere tin is mined as placer deposits, that is, deposits of sand and clay that contain tin that has been eroded from lodes and transported by streams. These are very low grade deposits and have to be processed over a number of stages.

◄ Cornish tin mines during the time when they were at full production during the last century.

Mining for lode tin

This diagram shows how hot fluids rising up from molten magma injected themselves into the rock surrounding the magma.

To recover this ore, deep shafts were sunk, following the near vertical directions of the deposits. The veins were typically very narrow, perhaps no more than a metre wide, something like sheets of material. The network of shafts and horizontal tunnels (adits) developed in a mine were designed to reach these mineral sheets from a number of places.

Lode ore contains a variety of impurities including tungsten, sulphur and arsenic, most of which can be removed by a stage of roasting before final smelting. Roasting arsenic-containing ores was a hazardous process.

igneous rock: a rock that has solidified from molten rock, either volcanic lava on the Earth's surface or magma deep underground. In either case the rock develops a network of interlocking crystals.

lode: a deposit in which a number of veins of a metal found close together.

magma: the molten rock that forms a balloon-shaped chamber in the rock below a volcano. It is fed by rock moving upwards from below the crust.

metamorphic rock: formed either from igneous or sedimentary rocks, by heat and or pressure. Metamorphic rocks form deep inside mountains during periods of mountain building. They result from the remelting of rocks during which process crystals are able to grow. Metamorphic rocks often show signs of banding and partial melting.

oxide: a compound that includes oxygen and one other element.

reduction: the removal of oxygen from a substance.

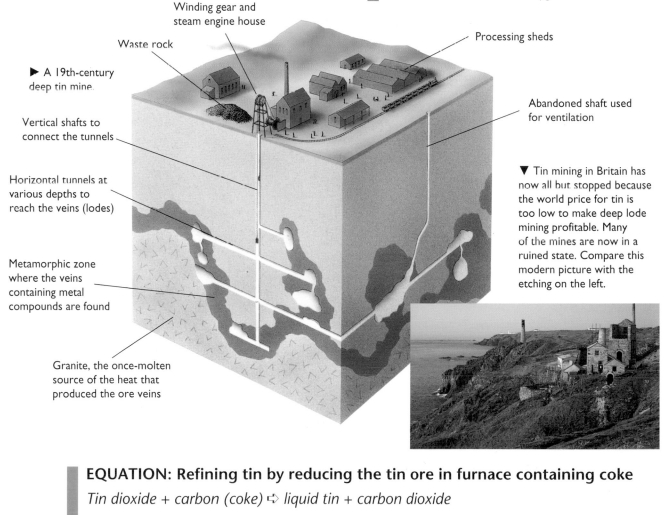

Winding gear and steam engine house

Waste rock

Processing sheds

▶ A 19th-century deep tin mine.

Abandoned shaft used for ventilation

Vertical shafts to connect the tunnels

Horizontal tunnels at various depths to reach the veins (lodes)

▼ Tin mining in Britain has now all but stopped because the world price for tin is too low to make deep lode mining profitable. Many of the mines are now in a ruined state. Compare this modern picture with the etching on the left.

Metamorphic zone where the veins containing metal compounds are found

Granite, the once-molten source of the heat that produced the ore veins

---

EQUATION: Refining tin by reducing the tin ore in furnace containing coke

Tin dioxide + carbon (coke) ⇨ liquid tin + carbon dioxide

$$SnO_2(s) \quad + \quad C(s) \quad \Rightarrow \quad Sn(s) \quad + \quad CO_2(g)$$

Placer tin mining

Four-fifths of the world's tin is obtained from placer deposits. Most placer deposits use high-pressure hoses to wash out the tin from deposits, or they use some form of excavator to dredge it from under the sea.

Placer deposits contain very low concentrations of tin which must be concentrated before they can be refined. Tin ore is a dense material and so the first stage of concentration is to wash it over ridges called riffles in a sluice box (see the picture on this page). The tin ore settles out and the other waste materials are washed away.

Unlike many other concentration techniques (such as that for lead shown on page 10), a relatively simple method such as sluicing concentrates the ore so that about three-quarters is tin oxide. This allows the placer ore to be smelted directly.

After smelting, the final impurities are removed from the tin by remelting and sometimes also by electrolysis. In the remelting method advantage is taken of the low melting point of tin. As the impure tin is heated, it melts and can be run off before the impurities begin to melt.

▶ Tin mining using placer deposits in Thailand. The working of such large deposits can cause great environmental damage unless care is taken with the waste materials.

Also...

The major tin-producing countries are Malaysia, Indonesia, Bolivia, Thailand, Russia, China and Brazil. The demand for tin has dropped in recent years as aluminium cans have supplanted steel cans. This change in use of metals has had a very severe impact on some developing world nations, depriving them of one of their sources of income.

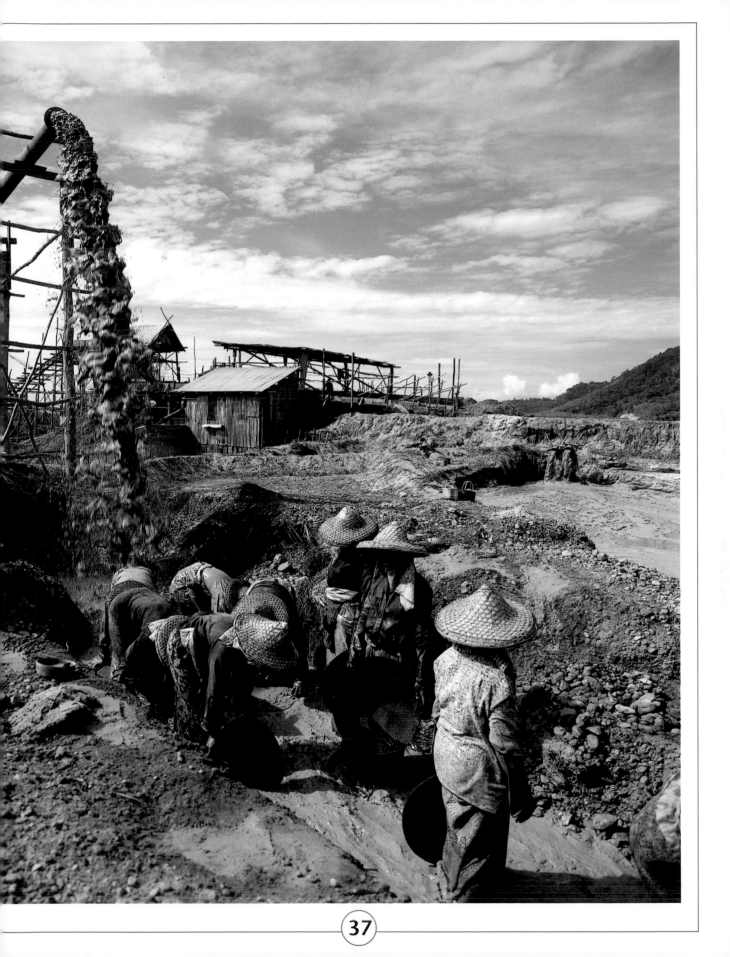

Tin plating

As soon as tin metal is exposed to the air, it develops a strong, gastight coating that prevents it from being corroded further. Although iron is next to tin in the reactivity series, the oxide coating that develops on iron is porous, which is why iron rusts. Thus tin (oxide) behaves very differently from iron. The low melting point of tin also means that it is easy to run sheet steel through a bath of tin without any fear that the steel might melt. These twin advantages have given tin its major use: as a way of protecting steel.

▲ Perhaps the world's largest tin object, the Margate Elephant, in New Jersey, USA, is as high as a six storey building and weighs 90 tonnes. It was built to attract tourists; its frame is made of wood, but its skin is made of tin to prevent corrosion in this exposed seaside environment.

▼ Tin plating is produced by pulling sheet steel through a bath of molten tin. This produces a coating on both sides of the steel.

Food containers

The advantage of tin is best seen in its use for food containers. Many foods contain organic acids. However, tin remains unreactive and so can be used for the inside of food cans because the food will not corrode the can. Tin plating was developed as a means of preserving moist food so that it could be taken to military forces when on active service. The French were among the first to try to invent a preserving process. In 1795 a Parisian chef, Nicolas Appert, invented the canning process. However, as tin-coated steel had not at that time been invented, his process actually involved preserving cooked food in bottles.

Bottles are not a good means of carrying food because of the risk of breakage. A new material had to be found. The tin-coated metal can was invented in Britain in 1810, and it was at this time that the word canning came into general use.

The value of the tin can was that the strength and cheapness of thin steel could be complemented by the corrosion resistance of soft tin. Plating the steel (which had to be done both inside and out) prevented the steel from rusting. The steel plate was curled into a cylinder and the seam made tight using a solder join. The top and bottom of the can could be fixed on with a press.

corrosive: a substance, either an acid or an alkali, that *rapidly* attacks a wide range of other substances.

plating: adding a thin coat of one material to another to make it resistant to corrosion.

REACTIVITY SERIES	
Element	*Reactivity*
potassium	*most reactive*
sodium	
calcium	
magnesium	
aluminium	
manganese	
chromium	
zinc	
iron	
cadmium	
tin	
lead	
copper	
mercury	
silver	
gold	
platinum	*least reactive*

◄ Metals can be placed in order of how vigorously they react in a list called a reactivity series. Tin is below iron in the reactivity series. Therefore exposed iron will always corrode when placed in a damp environment with tin. This is why steel cans rust badly as soon as they become scratched. It is not just the iron rusting; it is also encouraged to rust faster by the presence of the tin.

The disadvantage of using tin plate

Tin is plated onto iron or steel to protect it from corrosion, but it is only good if it remains unscratched. The reason you often see rusting cans is that tin is below steel (iron) in the reactivity series, so that when the steel is exposed by scratching, the steel corrodes at the expense of the tin, and the tin doesn't protect the steel at all. This is an important reason why aluminium containers have replaced tin-plated ones in recent years.

► A nail with a small tin collar has been placed in water. After a few days the nail is heavily rusted, but the tin is unaffected. This shows that tin speeds up the corrosion of steel and shows why tin-plated objects are only protected provided the tin remains unscratched.

Also...

Tin-plating is not the ideal way of protecting food. However, it was the best way for 150 years given the technology available until recently. Thus tin cans dominated the food-preservation market for 150 years because aluminium could not be produced cheaply.

Pewter and bronze

Pewter was originally an alloy of 40% lead and 60% tin. Because of the toxic effects of lead, modern pewter mostly contains tin with about 8% antimony as a hardening agent.

Pewter, which is a dark grey colour, was first made by the Romans, who used the tin from mines in Cornwall, England. As a result, England became an important centre for pewter.

Because it is quite soft, pewter was easily worked both by hammering and by turning on a lathe. It was used until the last century for all manner of goods, from utensils for the kitchen to goblets and chalices used in churches and monasteries. Pewter found its way to many parts of the world, and much of the antique pewter found in North America and Australasia was made in England.

Since the 19th century, when pewter was replaced by more durable metals and ceramics, pewter has mainly been used for decoration.

▲ Pewter was widely used because it was soft and easily worked. However, the high lead content meant that when it was used in articles intended for food or drink, some lead could be dissolved and produce a health hazard. Thus all modern pewter is lead free.

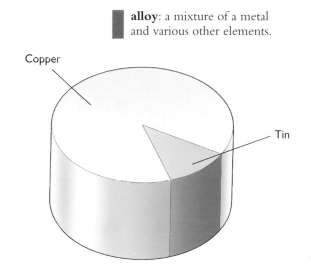

alloy: a mixture of a metal and various other elements.

Copper

Tin

▲ A bronze bell in Fort Sutter, Sacramento, California, USA.

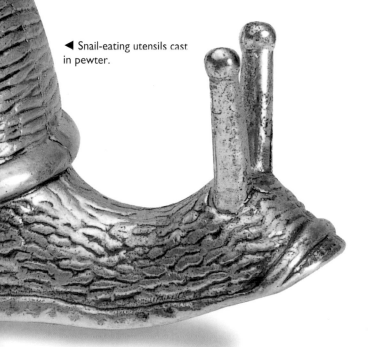

◀ Snail-eating utensils cast in pewter.

Bronze

Bronze is an alloy of copper and tin. Up to 10% of tin is normally used. Bronze has been used for decorative metal objects and also for coins. It was one of the earliest metal alloys made, giving rise to the first metal-working age, known as the Bronze Age, over three thousand years ago. Bronze Age people, however, did not know about alloying metals but used copper ores that contained tin impurities.

Bronzes are important because they are strong, they do not wear away easily (when used, for example, in gear wheels), and they resist saltwater corrosion. The rule is, the more tin there is in the alloy, the harder and more brittle it becomes. Bronzes with up to one-twelfth tin are used mainly for sheets, wire and coins; those with one-eighth tin are used mainly for gears, bearings and ships in exposed locations. When bronze contains up to one-fifth tin, it is used for gear wheels; and with up to one-quarter tin, bronze is used to make bells.

Also... special bronzes

A wide range of specialised bronzes are used, each with its own distinctive properties. These are produced by alloying with more elements. Phosphor bronze, for example, is commonly used for bearings in machines or engines where shafts continually rotate. Phosphor bronze is made by adding one-third of 1% of phosphorus to the alloy.

Silicon bronze is a strong metal, used in places where there is a great danger of corrosion, such as a chemical works. This alloy contains up to 3% silicon.

Solder

Soldering is a process of joining two pieces of metal together with a filler metal. Soldering metals have to be chosen to alloy with the metals they are joining and so provide a strong joint.

The composition of the solder is also designed to have a low melting point, so that it can be applied easily without damaging the metals being soldered.

Common solders are mixtures, or alloys, of lead and tin. Both lead and tin have low melting points, but a mixture of the two is stronger and has a lower melting point than either metal. Some materials, such as wires, are tin plated to make soldering easier.

The most common use of lead–tin soldering is to join copper wires or pipes.

To make a soldered joint
The use of lead solder relies on the way that a liquid will pull itself into a small space, called capillary action. The metal surfaces will only join if they are clean, so in general the tarnished surface is scraped away using steel wool, and the metal left bright and shiny. A material called a flux is then spread over the metals to be joined to prevent them from tarnishing again as they are heated.

For joining wires a soldering iron is used, and the solder has small rods of flux embedded in it. For soldering larger items, such as pipes, the heat has to be applied by a source such as a blowtorch. It is important that the heat is applied evenly so that the solder can run quickly and smoothly.

▼ Copper pipework is joined by solder. A flux is applied to the surfaces to be joined to prevent them from oxidising in the heat before the solder melts.

These ridges contain solder reservoirs. When the joint is heated with a blowtorch, the solder melts. A close fit ensures that capillary action causes the solder to flow evenly throughout the joint.

▶ The most commonly used solder consists of 50% tin and 50% lead. Solders containing a higher proportion of tin (say two-thirds tin, one-third lead) can be applied at a lower temperature but tend to be more brittle.

Lead solder and printed circuits

A printed circuit is an arrangement of electronic components, such as transistors and resistors, on a supporting board that is already printed with circuit wires.

The connections of the circuit are made with very thin layers of copper held in place by a glue. Too much heat and the glue will melt and the copper conductors will fall off the printed circuit board. The electronic components are equally delicate, and heat may easily crack them.

The key to making mass-produced printed circuits, found in everything from computers to washing machines, is to join the components to the copper on the board speedily and with as little heat as possible.

To do this the components are all pushed through holes in the board, and then the copper side is exposed. The protruding component wires are then soldered to the copper by passing them through an automatic soldering machine that applies just the right amount of solder for just the right length of time to make the correct joint before the board or components have the time to heat up too much.

capillary action: the tendency of a liquid to be sucked into small spaces, such as between objects and through narrow-pore tubes. The force to do this comes from surface tension.

flux: a material used to make it easier for a liquid to flow. A flux dissolves metal oxides and so prevents a metal from oxidising while being heated.

tarnish: a coating that develops as a result of the reaction between a metal and substances in the air. The most common form of tarnishing is a very thin transparent oxide coating.

▼ A section of circuit board showing the use of solder to connect components to the printed circuit.

Key facts about...

Lead

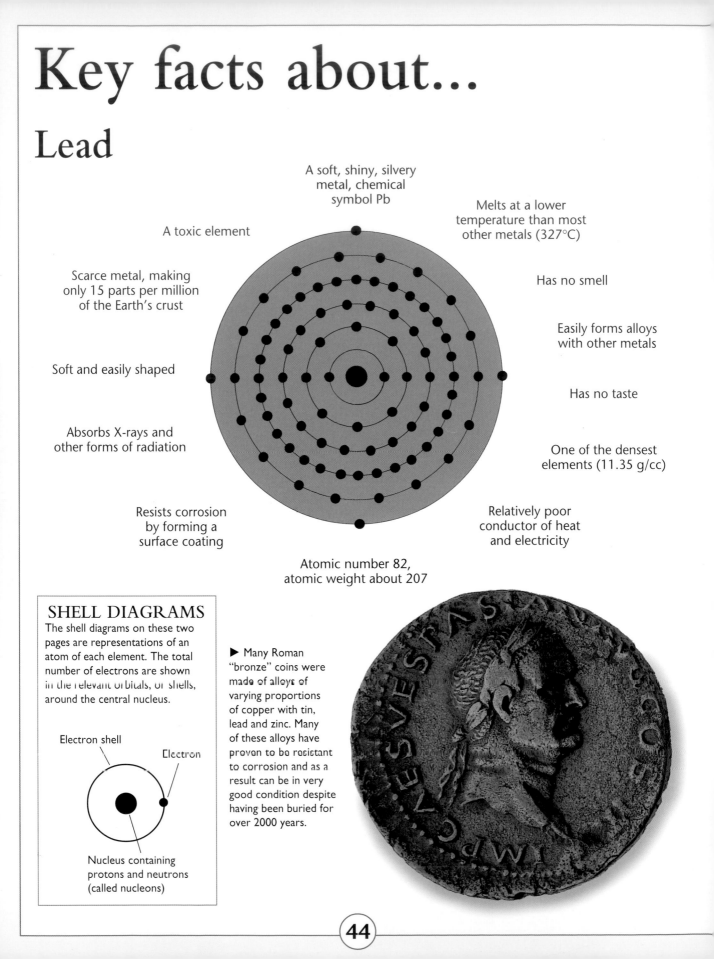

A soft, shiny, silvery metal, chemical symbol Pb

A toxic element

Melts at a lower temperature than most other metals (327°C)

Scarce metal, making only 15 parts per million of the Earth's crust

Has no smell

Easily forms alloys with other metals

Soft and easily shaped

Has no taste

Absorbs X-rays and other forms of radiation

One of the densest elements (11.35 g/cc)

Resists corrosion by forming a surface coating

Relatively poor conductor of heat and electricity

Atomic number 82, atomic weight about 207

SHELL DIAGRAMS

The shell diagrams on these two pages are representations of an atom of each element. The total number of electrons are shown in the relevant orbitals, or shells, around the central nucleus.

Electron shell

Electron

Nucleus containing protons and neutrons (called nucleons)

▶ Many Roman "bronze" coins were made of alloys of varying proportions of copper with tin, lead and zinc. Many of these alloys have proven to be resistant to corrosion and as a result can be in very good condition despite having been buried for over 2000 years.

Tin

A soft, shiny, silvery metal, chemical symbol Sn

Good conductor of heat and electricity

A weak metal that cannot be used alone

Has no smell

Does not tarnish when exposed to the air

Easily forms alloys with other metals

Has no taste

Density 7.3 g/cc

Melts at 232°C

Atomic number 50, atomic weight about 119

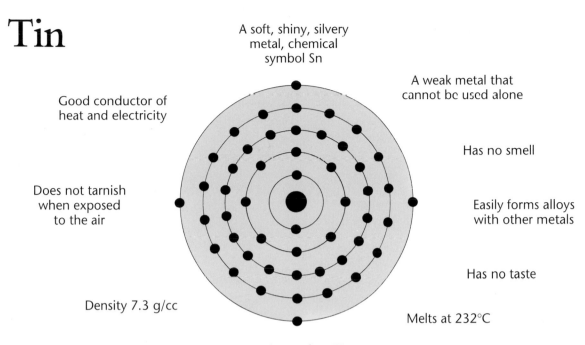

When a colourless solution of lead nitrate is added to a similarly colourless solution of potassium iodide, a thick yellow precipitate of lead iodide forms.

The Periodic Table

The Periodic Table sets out the relationships among the elements of the Universe. According to the Periodic Table, certain elements fall into groups. The pattern of these groups has, in the past, allowed scientists to predict elements that had not at that time been discovered. It can still be used today to predict the properties of unfamiliar elements.

The Periodic Table was first described by a Russian teacher, Dmitry Ivanovich Mendeleev, between 1869 and 1870. He was interested in writing a chemistry textbook, and wanted to show his students that there were certain patterns in the elements that had been discovered. So he set out the elements (of which there were 57 at the time) according to their known properties. On the assumption that there was pattern to the elements, he left blank spaces where elements seemed to be missing. Using this first version of the Periodic Table, he was able to predict in detail the chemical and physical properties of elements that had not yet been discovered. Other scientists began to look for the missing elements, and they soon found them.

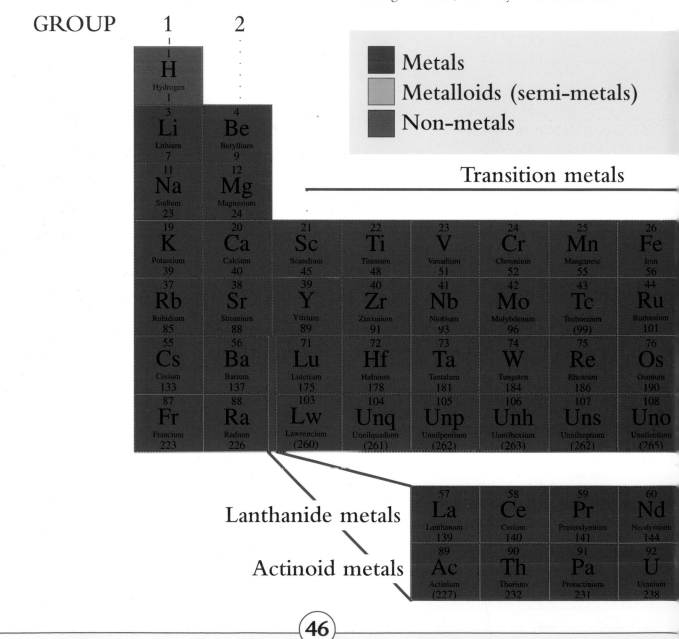

GROUP

Metals
Metalloids (semi-metals)
Non-metals

Transition metals

Lanthanide metals

Actinoid metals

Hydrogen did not seem to fit into the table, so he placed it in a box on its own. Otherwise the elements were all placed horizontally. When an element was reached with properties similar to the first one in the top row, a second row was started. By following this rule, similarities among the elements can be found by reading up and down. By reading across the rows, the elements progressively increase their atomic number. This number indicates the number of positively charged particles (protons) in the nucleus of each atom. This is also the number of negatively charged particles (electrons) in the atom.

The chemical properties of an element depend on the number of electrons in the outermost shell.

Atoms can form compounds by sharing electrons in their outermost shells. This explains why atoms with a full set of electrons (like helium, an inert gas) are unreactive, whereas atoms with an incomplete electron shell (such as chlorine) are very reactive. Elements can also combine by the complete transfer of electrons from metals to non-metals and the compounds formed contain ions.

Radioactive elements lose particles from their nucleus and electrons from their surrounding shells. As a result their atomic number changes and they become new elements.

Atomic (proton) number

13
Al — Symbol
Aluminium — Name
27

Approximate relative atomic mass
(Approximate atomic weight)

3	4	5	6	7	0
					2 He Helium 4
5 B Boron 11	6 C Carbon 12	7 N Nitrogen 14	8 O Oxygen 16	9 F Fluorine 19	10 Ne Neon 20
13 Al Aluminium 27	14 Si Silicon 28	15 P Phosphorus 31	16 S Sulphur 32	17 Cl Chlorine 35	18 Ar Argon 40

27 Co Cobalt 59	28 Ni Nickel 59	29 Cu Copper 64	30 Zn Zinc 65	31 Ga Gallium 70	32 Ge Germanium 73	33 As Arsenic 75	34 Se Selenium 79	35 Br Bromine 80	36 Kr Krypton 84
45 Rh Rhodium 103	46 Pd Palladium 106	47 Ag Silver 108	48 Cd Cadmium 112	49 In Indium 115	50 Sn Tin 119	51 Sb Antimony 122	52 Te Tellurium 128	53 I Iodine 127	54 Xe Xenon 131
77 Ir Iridium 192	78 Pt Platinum 195	79 Au Gold 197	80 Hg Mercury 201	81 Tl Thallium 204	82 Pb Lead 207	83 Bi Bismuth 209	84 Po Polonium (209)	85 At Astatine (210)	86 Rn Radon (222)
109 Une Unnilennium (266)									

61 Pm Promethium (145)	62 Sm Samarium 150	63 Eu Europium 152	64 Gd Gadolinium 157	65 Tb Terbium 159	66 Dy Dysprosium 163	67 Ho Holmium 165	68 Er Erbium 167	69 Tm Thulium 169	70 Yb Ytterbium 173
93 Np Neptunium (237)	94 Pu Plutonium (244)	95 Am Americium (243)	96 Cm Curium (247)	97 Bk Berkelium (247)	98 Cf Californium (251)	99 Es Einsteinium (252)	100 Fm Fermium (257)	101 Md Mendelevium (258)	102 No Nobelium (259)

Understanding equations

As you read through this book, you will notice that many pages contain equations using symbols. If you are not familiar with these symbols, read this page. Symbols make it easy for chemists to write out the reactions that are occurring in a way that allows a better understanding of the processes involved.

Symbols for the elements

The basis of the modern use of symbols for elements dates back to the 19th century. At this time a shorthand was developed using the first letter of the element wherever possible. Thus "O" stands for oxygen, "H" stands for hydrogen and so on. However, if we were to use only the first letter, then there could be some confusion. For example, nitrogen and nickel would both use the symbols N. To overcome this problem, many elements are symbolised using the first two letters of their full name, and the second letter is lowercase. Thus although nitrogen is N, nickel becomes Ni. Not all symbols come from the English name; many use the Latin name instead. This is why, for example, gold is not G but Au (for the Latin *aurum*) and sodium has the symbol Na, from the Latin *natrium*.

Compounds of elements are made by combining letters. Thus the molecule carbon

Written and symbolic equations

In this book, important chemical equations are briefly stated in words (these are called word equations), and are then shown in their symbolic form along with the states.

What reaction the equation illustrates

EQUATION: The formation of calcium hydroxide

Word equation —— *Calcium oxide + water ⇨ calcium hydroxide*

Symbol equation —— $CaO(s)$ + $H_2O(l)$ ⇨ $Ca(OH)_2(aq)$
                                                        heated

Sometimes you will find additional descriptions below the symbolic equation.

Symbol showing the state: *s* is for solid, *l* is for liquid, *g* is for gas and *aq* is for aqueous.

Diagrams

Some of the equations are shown as graphic representations.

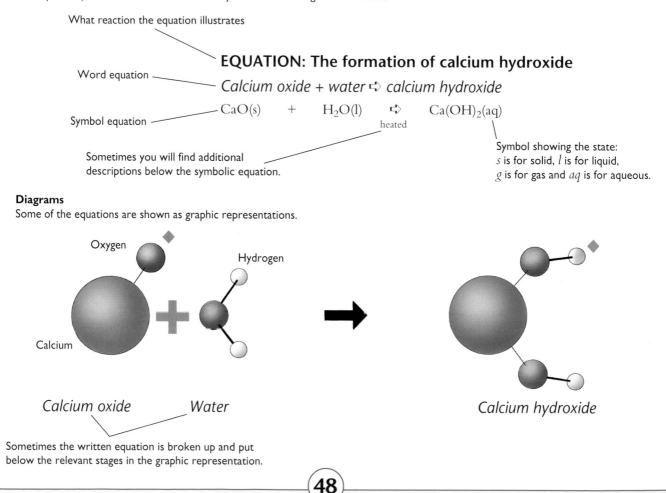

Oxygen

Hydrogen

Calcium

Calcium oxide   *Water*   *Calcium hydroxide*

Sometimes the written equation is broken up and put below the relevant stages in the graphic representation.

monoxide is CO. By using lowercase letters for the second letter of an element, it is possible to show that cobalt, symbol Co, is not the same as the molecule carbon monoxide, CO.

However, the letters can be made to do much more than this. In many molecules, atoms combine in unequal numbers. So, for example, carbon dioxide has one atom of carbon for every two of oxygen. This is shown by using the number 2 beside the oxygen, and the symbol becomes CO_2.

In practice, some groups of atoms combine as a unit with other substances. Thus, for example, calcium bicarbonate (one of the compounds used in some antacid pills) is written $Ca(HCO_3)_2$. This shows that the part of the substance inside the brackets reacts as a unit and the "2" outside the brackets shows the presence of two such units.

Some substances attract water molecules to themselves. To show this a dot is used. Thus the blue form of copper sulphate is written $CuSO_4.5H_2O$. In this case five molecules of water attract to one of copper sulphate.

When you see the dot, you know that this water can be driven off by heating; it is part of the crystal structure.

In a reaction substances change by rearranging the combinations of atoms. The way they change is shown by using the chemical symbols, placing those that will react (the starting materials, or reactants) on the left and the products of the reaction on the right. Between the two, chemists use an arrow to show which way the reaction is occurring.

It is possible to describe a reaction in words. This gives word equations, which are given throughout this book. However, it is easier to understand what is happening by using an equation containing symbols. These are also given in many places. They are not given when the equations are very complex.

In any equation both sides balance; that is, there must be an equal number of like atoms on both sides of the arrow. When you try to write down reactions, you, too, must balance your equation; you cannot have a few atoms left over at the end!

The symbols in brackets are abbreviations for the physical state of each substance taking part, so that (s) is used for solid, (l) for liquid, (g) for gas and (aq) for an aqueous solution, that is, a solution of a substance dissolved in water.

Atoms and ions
Each sphere represents a particle of an element. A particle can be an atom or an ion. Each atom or ion is associated with other atoms or ions through bonds – forces of attraction. The size of the particles and the nature of the bonds can be extremely important in determining the nature of the reaction or the properties of the compound.

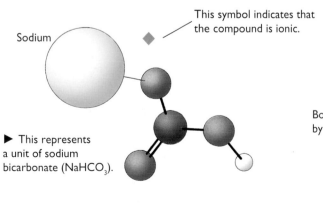

Sodium

This symbol indicates that the compound is ionic.

▶ This represents a unit of sodium bicarbonate ($NaHCO_3$).

The term "unit" is sometimes used to simplify the representation of a combination of ions.

Chemical symbols, equations and diagrams
The arrangement of any molecule or compound can be shown in one of the two ways shown below, depending on which gives the clearer picture. The left-hand diagram is called a ball-and-stick diagram because it uses rods and spheres to show the structure of the material. This example shows water, H_2O. There are two hydrogen atoms and one oxygen atom.

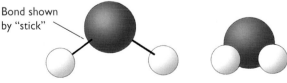

Bond shown by "stick"

Colours too
The colours of each of the particles help differentiate the elements involved. The diagram can then be matched to the written and symbolic equation given with the diagram. In the case above, oxygen is red and hydrogen is grey.

Glossary of technical terms

absorb: to soak up a substance. Compare to adsorb.

acetone: a petroleum-based solvent.

acid: compounds containing hydrogen which can attack and dissolve many substances. Acids are described as weak or strong, dilute or concentrated, mineral or organic.

acidity: a general term for the strength of an acid in a solution.

acid rain: rain that is contaminated by acid gases such as sulphur dioxide and nitrogen oxides released by pollution.

adsorb/adsorption: to "collect" gas molecules or other particles on to the *surface* of a substance. They are not chemically combined and can be removed. (The process is called "adsorption".) Compare to absorb.

alchemy: the traditional "art" of working with chemicals that prevailed through the Middle Ages. One of the main challenges of alchemy was to make gold from lead. Alchemy faded away as scientific chemistry was developed in the 17th century.

alkali: a base in solution.

alkaline: the opposite of acidic. Alkalis are bases that dissolve, and alkaline materials are called basic materials. Solutions of alkalis have a pH greater than 7.0 because they contain relatively few hydrogen ions.

alloy: a mixture of a metal and various other elements.

alpha particle: a stable combination of two protons and two neutrons, which is ejected from the nucleus of a radioactive atom as it decays. An alpha particle is also the nucleus of the atom of helium. If it captures two electrons it can become a neutral helium atom.

amalgam: a liquid alloy of mercury with another metal.

amino acid: amino acids are organic compounds that are the building blocks for the proteins in the body.

amorphous: a solid in which the atoms are not arranged regularly (i.e. "glassy"). Compare with crystalline.

amphoteric: a metal that will react with both acids and alkalis.

anhydrous: a substance from which water has been removed by heating. Many hydrated salts are crystalline. When they are heated and the water is driven off, the material changes to an anhydrous powder.

anion: a negatively charged atom or group of atoms.

anode: the negative terminal of a battery or the positive electrode of an electrolysis cell.

anodising: a process that uses the effect of electrolysis to make a surface corrosion-resistant.

antacid: a common name for any compound that reacts with stomach acid to neutralise it.

antioxidant: a substance that prevents oxidation of some other substance.

aqueous: a solid dissolved in water. Usually used as "aqueous solution".

atom: the smallest particle of an element.

atomic number: the number of electrons or the number of protons in an atom.

atomised: broken up into a very fine mist. The term is used in connection with sprays and engine fuel systems.

aurora: the "northern lights" and "southern lights" that show as coloured bands of light in the night sky at high latitudes. They are associated with the way cosmic rays interact with oxygen and nitrogen in the air.

basalt: an igneous rock with a low proportion of silica (usually below 55%). It has microscopically small crystals.

base: a compound that may be soapy to the touch and that can react with an acid in water to form a salt and water.

battery: a series of electrochemical cells.

bauxite: an ore of aluminium, of which about half is aluminium oxide.

becquerel: a unit of radiation equal to one nuclear disintegration per second.

beta particle: a form of radiation in which electrons are emitted from an atom as the nucleus breaks down.

bleach: a substance that removes stains from materials either by oxidising or reducing the staining compound.

boiling point: the temperature at which a liquid boils, changing from a liquid to a gas.

bond: chemical bonding is either a transfer or sharing of electrons by two or more atoms. There are a number of types of chemical bond, some very strong (such as covalent bonds), others weak (such as hydrogen bonds). Chemical bonds form because the linked molecule is more stable than the unlinked atoms from which it formed. For example, the hydrogen molecule (H_2) is more stable

than single atoms of hydrogen, which is why hydrogen gas is always found as molecules of two hydrogen atoms.

brass: a metal alloy principally of copper and zinc.

brazing: a form of soldering, in which brass is used as the joining metal.

brine: a solution of salt (sodium chloride) in water.

bronze: an alloy principally of copper and tin.

buffer: a chemistry term meaning a mixture of substances in solution that resists a change in the acidity or alkalinity of the solution.

capillary action: the tendency of a liquid to be sucked into small spaces, such as between objects and through narrow-pore tubes. The force to do this comes from surface tension.

catalyst: a substance that speeds up a chemical reaction but itself remains unaltered at the end of the reaction.

cathode: the positive terminal of a battery or the negative electrode of an electrolysis cell.

cathodic protection: the technique of making the object that is to be protected from corrosion into the cathode of a cell. For example, a material, such as steel, is protected by coupling it with a more reactive metal, such as magnesium. Steel forms the cathode and magnesium the anode. Zinc protects steel in the same way.

cation: a positively charged atom or group of atoms.

caustic: a substance that can cause burns if it touches the skin.

cell: a vessel containing two electrodes and an electrolyte that can act as an electrical conductor.

ceramic: a material based on clay minerals, which has been heated so that it has chemically hardened.

chalk: a pure form of calcium carbonate made of the crushed bodies of microscopic sea creatures, such as plankton and algae.

change of state: a change between one of the three states of matter, solid, liquid and gas.

chlorination: adding chlorine to a substance.

cladding: a surface sheet of material designed to protect other materials from corrosion.

clay: a microscopically small plate-like mineral that makes up the bulk of many soils. It has a sticky feel when wet.

combustion: the special case of oxidisation of a substance where a considerable amount of heat and usually light are given out. Combustion is often referred to as "burning".

compound: a chemical consisting of two or more elements chemically bonded together. Calcium atoms can combine with carbon atoms and oxygen atoms to make calcium carbonate, a compound of all three atoms.

condensation nuclei: microscopic particles of dust, salt and other materials suspended in the air, which attract water molecules.

conduction: (i) the exchange of heat (heat conduction) by contact with another object or (ii) allowing the flow of electrons (electrical conduction).

convection: the exchange of heat energy with the surroundings produced by the flow of a fluid due to being heated or cooled.

corrosion: the *slow* decay of a substance resulting from contact with gases and liquids in the environment. The term is often applied to metals. Rust is the corrosion of iron.

corrosive: a substance, either an acid or an alkali, that *rapidly* attacks a wide range of other substances.

cosmic rays: particles that fly through space and bombard all atoms on the Earth's surface. When they interact with the atmosphere they produce showers of secondary particles.

covalent bond: the most common form of strong chemical bonding, which occurs when two atoms *share* electrons.

cracking: breaking down complex molecules into simpler components. It is a term particularly used in oil refining.

crude oil: a chemical mixture of petroleum liquids. Crude oil forms the raw material for an oil refinery.

crystal: a substance that has grown freely so that it can develop external faces. Compare with crystalline, where the atoms are not free to form individual crystals and amorphous where the atoms are arranged irregularly.

crystalline: the organisation of atoms into a rigid "honeycomb-like" pattern without distinct crystal faces.

crystal systems: seven patterns or systems into which all of the world's crystals can be grouped. They are: cubic, hexagonal, rhombohedral, tetragonal, orthorhombic, monoclinic and triclinic.

cubic crystal system: groupings of crystals that look like cubes.

curie: a unit of radiation. The amount of radiation emitted by 1 g of radium each second. (The curie is equal to 37 billion becquerels.)

current: an electric current is produced by a flow of electrons through a conducting solid or ions through a conducting liquid.

decay (radioactive decay): the way that a radioactive element changes into another element because of loss of mass through radiation. For example uranium decays (changes) to lead.

decompose: to break down a substance (for example by heat or with the aid of a catalyst) into simpler components. In such a chemical reaction only one substance is involved.

dehydration: the removal of water from a substance by heating it, placing it in a dry atmosphere, or through the action of a drying agent.

density: the mass per unit volume (e.g. g/cc).

desertification: a process whereby a soil is allowed to become degraded to a state in which crops can no longer grow, i.e. desert-like. Chemical desertification is usually the result of contamination with halides because of poor irrigation practices.

detergent: a petroleum-based chemical that removes dirt.

diaphragm: a semipermeable membrane – a kind of ultra-fine mesh filter – that will allow only small ions to pass through. It is used in the electrolysis of brine.

diffusion: the slow mixing of one substance with another until the two substances are evenly mixed.

digestive tract: the system of the body that forms the pathway for food and its waste products. It begins at the mouth and includes the stomach and the intestines.

dilute acid: an acid whose concentration has been reduced by a large proportion of water.

diode: a semiconducting device that allows an electric current to flow in only one direction.

disinfectant: a chemical that kills bacteria and other microorganisms.

dissociate: to break apart. In the case of acids it means to break up forming hydrogen ions. This is an example of ionisation. Strong acids dissociate completely. Weak acids are not completely ionised and a solution of a weak acid has a relatively low concentration of hydrogen ions.

dissolve: to break down a substance in a solution without a resultant reaction.

distillation: the process of separating mixtures by condensing the vapours through cooling.

doping: adding metal atoms to a region of silicon to make it semiconducting.

dye: a coloured substance that will stick to another substance, so that both appear coloured.

electrode: a conductor that forms one terminal of a cell.

electrolysis: an electrical–chemical process that uses an electric current to cause the break up of a compound and the movement of metal ions in a solution. The process happens in many natural situations (as for example in rusting) and is also commonly used in industry for purifying (refining) metals or for plating metal objects with a fine, even metal coating.

electrolyte: a solution that conducts electricity.

electron: a tiny, negatively charged particle that is part of an atom. The flow of electrons through a solid material such as a wire produces an electric current.

electroplating: depositing a thin layer of a metal onto the surface of another substance using electrolysis.

element: a substance that cannot be decomposed into simpler substances by chemical means.

emulsion: tiny droplets of one substance dispersed in another. A common oil in water emulsion is milk. The tiny droplets in an emulsion tend to come together, so another stabilising substance is often needed to wrap the particles of grease and oil in a stable coat. Soaps and detergents are such agents. Photographic film is an example of a solid emulsion.

endothermic reaction: a reaction that takes heat from the surroundings. The reaction of carbon monoxide with a metal oxide is an example.

enzyme: organic catalysts in the form of proteins in the body that speed up chemical reactions. Every living cell contains hundreds of enzymes, which ensure that the processes of life continue. Should enzymes be made inoperative, such as through mercury poisoning, then death follows.

ester: organic compounds, formed by the reaction of an alcohol with an acid, which often have a fruity taste.

evaporation: the change of state of a liquid to a gas. Evaporation happens below the boiling point and is used as a method of separating out the materials in a solution.

exothermic reaction: a reaction that gives heat to the surroundings. Many oxidation reactions, for example, give out heat.

explosive: a substance which, when a shock is applied to it, decomposes very rapidly, releasing a very large amount of heat and creating a large volume of gases as a shock wave.

extrusion: forming a shape by pushing it through a die. For example, toothpaste is extruded through the cap (die) of the toothpaste tube.

fallout: radioactive particles that reach the ground from radioactive materials in the atmosphere.

fat: semi-solid energy-rich compounds derived from plants or animals and which are made of carbon, hydrogen and oxygen. Scientists call these esters.

feldspar: a mineral consisting of sheets of aluminium silicate. This is the mineral from which the clay in soils is made.

fertile: able to provide the nutrients needed for unrestricted plant growth.

filtration: the separation of a liquid from a solid using a membrane with small holes.

fission: the breakdown of the structure of an atom, popularly called "splitting the atom" because the atom is split into approximately two other nuclei. This is different from, for example, the small change that happens when radioactivity is emitted.

fixation of nitrogen: the processes that natural organisms, such as bacteria, use to turn the nitrogen of the air into ammonium compounds.

fixing: making solid and liquid nitrogen-containing compounds from nitrogen gas. The compounds that are formed can be used as fertilisers.

fluid: able to flow; either a liquid or a gas.

fluorescent: a substance that gives out visible light when struck by invisible waves such as ultraviolet rays.

flux: a material used to make it easier for a liquid to flow. A flux dissolves metal oxides and so prevents a metal from oxidising while being heated.

foam: a substance that is sufficiently gelatinous to be able to contain bubbles of gas. The gas bulks up the substance, making it behave as though it were semi-rigid.

fossil fuels: hydrocarbon compounds that have been formed from buried plant and animal remains. High pressures and temperatures lasting over millions of years are required. The fossil fuels are coal, oil and natural gas.

fraction: a group of similar components of a mixture. In the petroleum industry the light fractions of crude oil are those with the smallest molecules, while the medium and heavy fractions have larger molecules.

free radical: a very reactive atom or group with a "spare" electron.

freezing point: the temperature at which a substance changes from a liquid to a solid. It is the same temperature as the melting point.

fuel: a concentrated form of chemical energy. The main sources of fuels (called fossil fuels because they were formed by geological processes) are coal, crude oil and natural gas. Products include methane, propane and gasoline. The fuel for stars and space vehicles is hydrogen.

fuel rods: rods of uranium or other radioactive material used as a fuel in nuclear power stations.

fuming: an unstable liquid that gives off a gas. Very concentrated acid solutions are often fuming solutions.

fungicide: any chemical that is designed to kill fungi and control the spread of fungal spores.

fusion: combining atoms to form a heavier atom.

galvanising: applying a thin zinc coating to protect another metal.

gamma rays: waves of radiation produced as the nucleus of a radioactive element rearranges itself into a tighter cluster of protons and neutrons. Gamma rays carry enough energy to damage living cells.

gangue: the unwanted material in an ore.

gas: a form of matter in which the molecules form no definite shape and are free to move about to fill any vessel they are put in.

gelatinous: a term meaning made with water. Because a gelatinous precipitate is mostly water, it is of a similar density to water and will float or lie suspended in the liquid.

gelling agent: a semi-solid jelly-like substance.

gemstone: a wide range of minerals valued by people, both as crystals (such as emerald) and as decorative stones (such as agate). There is no single chemical formula for a gemstone.

glass: a transparent silicate without any crystal growth. It has a glassy lustre and breaks with a curved fracture. Note that some minerals have all these features and are therefore natural glasses. Household glass is a synthetic silicate.

glucose: the most common of the natural sugars. It occurs as the polymer known as cellulose, the fibre in plants. Starch is also a form of glucose. The breakdown of glucose provides the energy that animals need for life.

granite: an igneous rock with a high proportion of silica (usually over 65%). It has well-developed large crystals. The largest pink, grey or white crystals are feldspar.

Greenhouse Effect: an increase of the global air temperature as a result of heat released from burning fossil fuels being absorbed by carbon dioxide in the atmosphere.

gypsum: the name for calcium sulphate. It is commonly found as Plaster of Paris and wallboards.

half-life: the time it takes for the radiation coming from a sample of a radioactive element to decrease by half.

halide: a salt of one of the halogens (fluorine, chlorine, bromine and iodine).

halite: the mineral made of sodium chloride.

halogen: one of a group of elements including chlorine, bromine, iodine and fluorine.

heat-producing: see exothermic reaction.

high explosive: a form of explosive that will only work when it receives a shock from another explosive. High explosives are much more powerful than ordinary explosives. Gunpowder is not a high explosive.

hydrate: a solid compound in crystalline form that contains molecular water. Hydrates commonly form when a solution of a soluble salt is evaporated. The water that forms part of a hydrate crystal is known as the "water of crystallization". It can usually be removed by heating, leaving an anhydrous salt.

hydration: the absorption of water by a substance. Hydrated materials are not "wet" but remain firm, apparently dry, solids. In some cases, hydration makes the substance change colour, in many other cases there is no colour change, simply a change in volume.

hydrocarbon: a compound in which only hydrogen and carbon atoms are present. Most fuels are hydrocarbons, as is the simple plastic polyethene (known as polythene).

hydrogen bond: a type of attractive force that holds one molecule to another. It is one of the weaker forms of intermolecular attractive force.

hydrothermal: a process in which hot water is involved. It is usually used in the context of rock formation because hot water and other fluids sent outwards from liquid magmas are important carriers of metals and the minerals that form gemstones.

igneous rock: a rock that has solidified from molten rock, either volcanic lava on the Earth's surface or magma deep underground. In either case the rock develops a network of interlocking crystals.

incendiary: a substance designed to cause burning.

indicator: a substance or mixture of substances that change colour with acidity or alkalinity.

inert: nonreactive.

infra-red radiation: a form of light radiation where the wavelength of the waves is slightly longer than visible light. Most heat radiation is in the infra-red band.

insoluble: a substance that will not dissolve.

ion: an atom, or group of atoms, that has gained or lost one or more electrons and so developed an electrical charge. Ions behave differently from electrically neutral atoms and molecules. They can move in an electric field, and they can also bind strongly to solvent molecules such as water. Positively charged ions are called cations; negatively charged ions are called anions. Ions carry electrical current through solutions.

ionic bond: the form of bonding that occurs between two ions when the ions have opposite charges. Sodium cations bond with chloride anions to form common salt (NaCl) when a salty solution is evaporated. Ionic bonds are strong bonds except in the presence of a solvent.

ionise: to break up neutral molecules into oppositely charged ions or to convert atoms into ions by the loss of electrons.

ionisation: a process that creates ions.

irrigation: the application of water to fields to help plants grow during times when natural rainfall is sparse.

isotope: atoms that have the same number of protons in their nucleus, but which have different masses; for example, carbon-12 and carbon-14.

latent heat: the amount of heat that is absorbed or released during the process of changing state between gas, liquid or solid. For example, heat is absorbed when a substance melts and it is released again when the substance solidifies.

latex. (the Latin word for "liquid") a suspension of small polymer particles in water. The rubber that flows from a rubber tree is a natural latex. Some synthetic polymers are made as latexes, allowing polymerisation to take place in water.

lava: the material that flows from a volcano.

limestone: a form of calcium carbonate rock that is often formed of lime mud Most limestones are light grey and have abundant fossils.

liquid: a form of matter that has a fixed volume but no fixed shape.

lode: a deposit in which a number of veins of a metal found close together.

lustre: the shininess of a substance.

magma: the molten rock that forms a balloon-shaped chamber in the rock below a volcano. It is fed by rock moving upwards from below the crust.

marble: a form of limestone that has been "baked" while deep inside mountains. This has caused the limestone to melt and reform into small interlocking crystals, making marble harder than limestone.

mass: the amount of matter in an object. In everyday use, the word weight is often used to mean mass

melting point: the temperature at which a substance changes state from a solid to a liquid. It is the same as freezing point.

membrane: a thin flexible sheet. A semipermeable membrane has microscopic holes of a size that will selectively allow some ions and molecules to pass through but hold others back. It thus acts as a kind of sieve.

meniscus: the curved surface of a liquid that forms when it rises in a small bore, or capillary tube. The meniscus is convex (bulges upwards) for mercury and is concave (sags downwards) for water.

metal: a substance with a lustre, the ability to conduct heat and electricity and which is not brittle.

metallic bonding: a kind of bonding in which atoms reside in a "sea" of mobile electrons. This type of bonding allows metals to be good conductors and means that they are not brittle

metamorphic rock: formed either from igneous or sedimentary rocks, by heat and or pressure. Metamorphic rocks form deep inside mountains during periods of mountain building. They result from the remelting of rocks during which process crystals are able to grow. Metamorphic rocks often show signs of banding and partial melting.

micronutrient: an element that the body requires in small amounts. Another term is trace element.

mineral: a solid substance made of just one element or chemical compound. Calcite is a mineral because it consists only of calcium carbonate, halite is a mineral because it contains only sodium chloride, quartz is a mineral because it consists of only silicon dioxide.

mineral acid: an acid that does not contain carbon and that attacks minerals. Hydrochloric, sulphuric and nitric acids are the main mineral acids.

mineral-laden: a solution close to saturation.

mixture: a material that can be separated out into two or more substances using physical means.

molecule: a group of two or more atoms held together by chemical bonds.

monoclinic system: a grouping of crystals that look like double-ended chisel blades.

monomer: a building block of a larger chain molecule ("mono" means one, "mer" means part).

mordant: any chemical that allows dyes to stick to other substances.

native metal: a pure form of a metal, not combined as a compound. Native metal is more common in poorly reactive elements than in those that are very reactive

neutralisation: the reaction of acids and bases to produce a salt and water. The reaction causes hydrogen from the acid and hydroxide from the base to be changed to water. For

example, hydrochloric acid reacts with sodium hydroxide to form common salt and water. The term is more generally used for any reaction where the pH changes towards 7.0, which is the pH of a neutral solution.

neutron: a particle inside the nucleus of an atom that is neutral and has no charge.

noncombustible: a substance that will not burn.

noble metal: silver, gold, platinum, and mercury. These are the least reactive metals.

nuclear energy: the heat energy produced as part of the changes that take place in the core, or nucleus, of an element's atoms.

nuclear reactions: reactions that occur in the core, or nucleus of an atom.

nutrients: soluble ions that are essential to life.

octane: one of the substances contained in fuel.

ore: a rock containing enough of a useful substance to make mining it worthwhile.

organic acid: an acid containing carbon and hydrogen.

organic substance: a substance that contains carbon.

osmosis: a process where molecules of a liquid solvent move through a membrane (filter) from a region of low concentration to a region of high concentration of solute.

oxidation: a reaction in which the oxidising agent removes electrons. (Note that oxidising agents do not have to contain oxygen.)

oxide: a compound that includes oxygen and one other element.

oxidise: the process of gaining oxygen. This can be part of a controlled chemical reaction, or it can be the result of exposing a substance to the air, where oxidation (a form of corrosion) will occur slowly, perhaps over months or years.

oxidising agent: a substance that removes electrons from another substance (and therefore is itself reduced).

ozone: a form of oxygen whose molecules contain three atoms of oxygen. Ozone is regarded as a beneficial gas when high in the atmosphere because it blocks ultraviolet rays. It is a harmful gas when breathed in, so low level ozone, which is produced as part of city smog, is regarded as a form of pollution. The ozone layer is the uppermost part of the stratosphere.

pan: the name given to a shallow pond of liquid Pans are mainly used for separating solutions by evaporation.

patina: a surface coating that develops on metals and protects them from further corrosion.

percolate: to move slowly through the pores of a rock.

period: a row in the Periodic Table.

Periodic Table: a chart organising elements by atomic number and chemical properties into groups and periods.

pesticide: any chemical that is designed to control pests (unwanted organisms) that are harmful to plants or animals.

petroleum: a natural mixture of a range of gases, liquids and solids derived from the decomposed remains of plants and animals.

pH: a measure of the hydrogen ion concentration in a liquid. Neutral is pH 7.0; numbers greater than this are alkaline, smaller numbers are acidic.

phosphor: any material that glows when energized by ultraviolet or electron beams such as in fluorescent tubes and cathode ray tubes. Phosphors, such as phosphorus, emit light after the source of excitation is cut off. This is why they glow in the dark. By contrast, fluorescors, such as fluorite, emit light only while they are being excited by ultraviolet light or an electron beam.

photon: a parcel of light energy.

photosynthesis: the process by which plants use the energy of the Sun to make the compounds they need for life. In photosynthesis, six molecules of carbon dioxide from the air combine with six molecules of water, forming one molecule of glucose (sugar) and releasing six molecules of oxygen back into the atmosphere.

pigment: any solid material used to give a liquid a colour.

placer deposit: a kind of ore body made of a sediment that contains fragments of gold ore eroded from a mother lode and transported by rivers and/or ocean currents.

plastic (material): a carbon-based material consisting of long chains (polymers) of simple molecules. The word plastic is commonly restricted to synthetic polymers.

plastic (property): a material is plastic if it can be made to change shape easily. Plastic materials will remain in the new shape. (Compare with elastic, a property where a material goes back to its original shape.)

plating: adding a thin coat of one material to another to make it resistant to corrosion.

playa: a dried-up lake bed that is covered with salt deposits. From the Spanish word for beach.

poison gas: a form of gas that is used intentionally to produce widespread injury and death. (Many gases are poisonous, which is why many chemical reactions are performed in laboratory fume chambers, but they are a byproduct of a reaction and not intended to cause harm.)

polymer: a compound that is made of long chains by combining molecules (called monomers) as repeating units. ("Poly" means many, "mer" means part).

polymerisation: a chemical reaction in which large numbers of similar molecules arrange themselves into large molecules, usually long chains. This process usually happens when there is a suitable catalyst present. For example, ethene reacts to form polythene in the presence of certain catalysts.

porous: a material containing many small holes or cracks. Quite often the pores are connected, and liquids, such as water or oil, can move through them.

precious metal: silver, gold, platinum, iridium, and palladium. Each is prized for its rarity. This category is the equivalent of precious stones, or gemstones, for minerals.

precipitate: tiny solid particles formed as a result of a chemical reaction between two liquids or gases.

preservative: a substance that prevents the natural organic decay processes from occurring. Many substances can be used safely for this purpose, including sulphites and nitrogen gas.

product: a substance produced by a chemical reaction.

protein: molecules that help to build tissue and bone and therefore make new body cells. Proteins contain amino acids.

proton: a positively charged particle in the nucleus of an atom that balances out the charge of the surrounding electrons

pyrite: "mineral of fire". This name comes from the fact that pyrite (iron sulphide) will give off sparks if struck with a stone.

pyrometallurgy: refining a metal from its ore using heat. A blast furnace or smelter is the main equipment used.

radiation: the exchange of energy with the surroundings through the transmission of waves or particles of energy. Radiation is a form of energy transfer that can happen through space; no intervening medium is required (as would be the case for conduction and convection).

radioactive: a material that emits radiation or particles from the nucleus of its atoms.

radioactive decay: a change in a radioactive element due to loss of mass through radiation. For example uranium decays (changes) to lead.

radioisotope: a shortened version of the phrase radioactive isotope.

radiotracer: a radioactive isotope that is added to a stable, nonradioactive material in order to trace how it moves and its concentration.

reaction: the recombination of two substances using parts of each substance to produce new substances.

reactivity: the tendency of a substance to react with other substances. The term is most widely used in comparing the reactivity of metals. Metals are arranged in a reactivity series.

reagent: a starting material for a reaction.

recycling: the reuse of a material to save the time and energy required to extract new material from the Earth and to conserve non-renewable resources.

redox reaction: a reaction that involves reduction and oxidation.

reducing agent: a substance that gives electrons to another substance. Carbon monoxide is a reducing agent when passed over copper oxide, turning it to copper and producing carbon dioxide gas. Similarly, iron oxide is reduced to iron in a blast furnace. Sulphur dioxide is a reducing agent, used for bleaching bread.

reduction: the removal of oxygen from a substance. See also: oxidation.

refining: separating a mixture into the simpler substances of which it is made. In the case of a rock, it means the extraction of the metal that is mixed up in the rock. In the case of oil it means separating out the fractions of which it is made.

refractive index: the property of a transparent material that controls the angle at which total internal reflection will occur. The greater the refractive index, the more reflective the material will be.

resin: natural or synthetic polymers that can be moulded into solid objects or spun into thread.

rust: the corrosion of iron and steel.

saline: a solution in which most of the dissolved matter is sodium chloride (common salt).

salinisation: the concentration of salts, especially sodium chloride, in the upper layers of a soil due to poor methods of irrigation.

salts: compounds, often involving a metal, that are the reaction products of acids and bases. (Note "salt" is also the common word for sodium chloride, common salt or table salt.)

saponification: the term for a reaction between a fat and a base that produces a soap.

saturated: a state where a liquid can hold no more of a substance. If any more of the substance is added, it will not dissolve.

saturated solution: a solution that holds the maximum possible amount of dissolved material. The amount of material in solution varies with the temperature; cold solutions can hold less dissolved solid material than hot solutions. Gases are more soluble in cold liquids than hot liquids.

sediment: material that settles out at the bottom of a liquid when it is still.

semiconductor: a material of intermediate conductivity. Semiconductor devices often use silicon when they are made as part of diodes, transistors or integrated circuits.

semipermeable membrane: a thin (membrane) of material that acts as a fine sieve, allowing small molecules to pass, but holding large molecules back.

silicate: a compound containing silicon and oxygen (known as silica).

sintering: a process that happens at moderately high temperatures in some compounds. Grains begin to fuse together even through they do not melt. The most widespread example of sintering happens during the firing of clays to make ceramics.

slag: a mixture of substances that are waste products of a furnace. Most slags are composed mainly of silicates.

smelting: roasting a substance in order to extract the metal contained in it.

smog: a mixture of smoke and fog. The term is used to describe city fogs in which there is a large proportion of particulate matter (tiny pieces of carbon from exhausts) and also a high concentration of sulphur and nitrogen gases and probably ozone.

soldering: joining together two pieces of metal using solder, an alloy with a low melting point.

solid: a form of matter where a substance has a definite shape.

soluble: a substance that will readily dissolve in a solvent.

solute: the substance that dissolves in a solution (e.g. sodium chloride in salt water).

solution: a mixture of a liquid and at least one other substance (e.g. salt water). Mixtures can be separated out by physical means, for example by evaporation and cooling.

solvent: the main substance in a solution (e.g. water in salt water).

spontaneous combustion: the effect of a very reactive material beginning to oxidise very quickly and bursting into flame.

stable: able to exist without changing into another substance.

stratosphere: the part of the Earth's atmosphere that lies immediately above the region in which clouds form. It occurs between 12 and 50 km above the Earth's surface.

strong acid: an acid that has completely dissociated (ionised) in water. Mineral acids are strong acids.

sublimation: the change of a substance from solid to gas, or vica versa, without going through a liquid phase.

substance: a type of material, including mixtures.

sulphate: a compound that includes sulphur and oxygen, for example, calcium sulphate or gypsum.

sulphide: a sulphur compound that contains no oxygen.

sulphite: a sulphur compound that contains less oxygen than a sulphate.

surface tension: the force that operates on the surface of a liquid, which makes it act as though it were covered with an invisible elastic film.

suspension: tiny particles suspended in a liquid.

synthetic: does not occur naturally, but has to be manufactured.

tarnish: a coating that develops as a result of the reaction between a metal and substances in the air. The most common form of tarnishing is a very thin transparent oxide coating.

thermonuclear reactions: reactions that occur within atoms due to fusion, releasing an immensely concentrated amount of energy.

thermoplastic: a plastic that will soften, can repeatedly be moulded it into shape on heating and will set into the moulded shape as it cools.

thermoset: a plastic that will set into a moulded shape as it cools, but which cannot be made soft by reheating.

titration: a process of dripping one liquid into another in order to find out the amount needed to cause a neutral solution. An indicator is used to signal change.

toxic: poisonous enough to cause death.

translucent: almost transparent.

transmutation: the change of one element into another.

vapour: the gaseous form of a substance that is normally a liquid. For example, water vapour is the gaseous form of liquid water.

vein: a mineral deposit different from, and usually cutting across, the surrounding rocks. Most mineral and metal-bearing veins are deposits filling fractures. The veins were filled by hot, mineral-rich waters rising upwards from liquid volcanic magma. They are important sources of many metals, such as silver and gold, and also minerals such as gemstones. Veins are usually narrow, and were best suited to hand-mining. They are less exploited in the modern machine age.

viscous: slow moving, syrupy. A liquid that has a low viscosity is said to be mobile.

vitreous: glass-like.

volatile: readily forms a gas.

vulcanisation: forming cross-links between polymer chains to increase the strength of the whole polymer. Rubbers are vulcanised using sulphur when making tyres and other strong materials.

weak acid: an acid that has only partly dissociated (ionised) in water. Most organic acids are weak acids.

weather: a term used by Earth scientists and derived from "weathering", meaning to react with water and gases of the environment.

weathering: the slow natural processes that break down rocks and reduce them to small fragments either by mechanical or chemical means.

welding: fusing two pieces of metal together using heat.

X-rays: a form of very short wave radiation.

Index

04510